Rao, Schott

Understanding Plastics Engineering Calculations

Natti S. Rao
Nick R. Schott

Understanding
Plastics Engineering
Calculations

Hands-on Examples and Case Studies

HANSER

The Authors:
Dr. Natti S. Rao, 327 Route 21c, Ghent, NY 12075, USA
Dr. Nick R. Schott, University of Massachusetts at Lowell, 1 University Avenue, Lowell, MA 01854, USA

Distributed in North and South America by:
Hanser Publications
6915 Valley Avenue, Cincinnati, Ohio 45244-3029, USA
Fax: (513) 527-8801
Phone: (513) 527-8977
www.hanserpublications.com

Distributed in all other countries by
Carl Hanser Verlag
Postfach 86 04 20, 81631 München, Germany
Fax: +49 (89) 98 48 09
www.hanser.de

The use of general descriptive names, trademarks, etc., in this publication, even if the former are not especially identified, is not to be taken as a sign that such names, as understood by the Trade Marks and Merchandise Marks Act, may accordingly be used freely by anyone.
While the advice and information in this book are believed to be true and accurate at the date of going to press, neither the authors nor the editors nor the publisher can accept any legal responsibility for any errors or omissions that may be made. The publisher makes no warranty, express or implied, with respect to the material contained herein.

Library of Congress Cataloging-in-Publication Data

Rao, Natti S.
  Understanding plastics engineering calculations / Natti S. Rao, Nick R. Schott.
     p. cm.
  Includes bibliographical references and index.
  ISBN 978-1-56990-509-8 (hardcover) -- ISBN 1-56990-509-6 (hardcover) -- ISBN 978-3-446-42278-0 (hardcover)
1. Plastics--Molding--Mathematical models. I. Schott, Nick R., 1939- II. Title.
  TP1150.R36 2012
  624.1'8923--dc23
                        2011051289

Bibliografische Information Der Deutschen Bibliothek
Die Deutsche Bibliothek verzeichnet diese Publikation in der Deutschen Nationalbibliografie;
detaillierte bibliografische Daten sind im Internet über <http://dnb.d-nb.de> abrufbar.

ISBN 978-3-446-42278-0
E-Book-ISBN 978-3-446-43149-2

© Carl Hanser Verlag, Munich 2012
Production Management: Steffen Jörg
Coverconcept: Marc Müller-Bremer, www.rebranding.de, München
Coverdesign: Stephan Rönigk
Typeset: Manuela Treindl, Fürth
Printed and bound by Kösel, Krugzell
Printed in Germany

Table of Contents

Preface IX

1 Rheological Properties of Molten Polymers 1
   1.1 Polymer Melt Flow 1
      1.1.1 Apparent Shear Rate 2
      1.1.2 Apparent Viscosity 3
      1.1.3 Power Law of Ostwald and De Waele 3
      1.1.4 Viscosity Formula of Klein 4
      1.1.5 Resin Characterization by Power Law Exponent 5
   1.2 Melt Flow Index 7
   1.3 Relationship between Flow Rate and Pressure Drop 8
   1.4 Shear Rates for Extrusion Dies 10
   References 13

2 Thermal Properties of Solid and Molten Polymers 15
   2.1 Specific Volume 15
   2.2 Specific Heat 18
   2.3 Thermal Expansion Coefficient 19
   2.4 Enthalpy 20
   2.5 Thermal Conductivity 22
   2.6 Thermal Diffusivity 23
   2.7 Coefficient of Heat Penetration 24
   2.8 Heat Deflection Temperature 25
   2.9 Vicat Softening Point 26
   References 28

3 Heat Transfer in Plastics Processing 29
   3.1 Steady State Conduction 29
      3.1.1 Plane Wall 30
      3.1.2 Cylinder 31

3.1.3   Hollow Sphere 31
3.1.4   Sphere 32
3.1.5   Heat Conduction in Composite Walls 32
3.1.6   Overall Heat Transfer through Composite Walls 36
3.2   Unsteady State Conduction 37
3.2.1   Temperature Distribution in One-Dimensional Solids 38
3.2.2   Thermal Contact Temperature 45
3.3   Heat Conduction with Dissipation 47
3.4   Dimensionless Groups 48
3.5   Heat Transfer by Convection 51
3.6   Heat Transfer by Radiation 53
3.7   Dielectric Heating 57
3.8   Fick's Law of Diffusion 59
3.8.1   Permeability 59
3.8.2   Absorption and Desorption 60
3.9   Case Study: Analyzing Air Gap Dynamics in Extrusion Coating by Means of
       Dimensional Analysis 61
3.9.1   Heat Transfer Between the Film and the Surrounding Air 62
3.9.2   Chemical Kinetics 63
3.9.3   Evaluation of the Experiments 65
References 66

4   Analytical Procedures for Troubleshooting Extrusion Screws 67
4.1   Three-Zone Screw 68
4.1.1   Extruder Output 69
4.1.2   Feed Zone 69
4.1.3   Metering Zone (Melt Zone) 71
4.1.4   Practical Design of 3-Zone Screws 78
4.2   Melting of Solids 83
4.2.1   Thickness of Melt Film 83
4.2.2   Melting Profile 87
4.2.3   Melt Temperature 91
4.2.4   Melt Pressure 92
4.2.5   Heat Transfer between the Melt and the Barrel 93
4.2.6   Screw Power 95
4.2.7   Temperature Fluctuation of the Melt 98
4.2.8   Pressure Fluctuation 99
4.2.9   Extrusion Screw Simulations 99
4.2.10   Mechanical Design of Extrusion Screws 107
References 110

5 Analytical Procedures for Troubleshooting Extrusion Dies 111
   5.1 Calculation of Pressure Drop 112
      5.1.1 Effect of Die Geometry on Pressure Drop 113
      5.1.2 Shear Rate in Die Channels 114
      5.1.3 General Relationship for Pressure Drop in Any Given Channel Geometry .. 114
      5.1.4 Examples for Calculating Pressure Drop in the Die Channels of Different Shapes 115
      5.1.5 Temperature Rise and Residence Time 123
   5.2 Spider Dies 124
   5.3 Spiral Dies 129
   5.4 Adapting Die Design to Avoid Melt Fracture 130
      5.4.1 Pelletizer Dies 132
      5.4.2 Blow Molding Dies 132
      5.4.3 Summary of the Die Design Procedures 134
   5.5 Flat Dies 135
   5.6 An Easily Applicable Method of Designing Screen Packs for Extruders 137
   5.7 Parametrical Studies 143
      5.7.1 Pipe Extrusion 143
      5.7.2 Blown Film 146
      5.7.3 Thermoforming 149
   References 152

6 Analytical Procedures for Troubleshooting Injection Molding 153
   6.1 Effect of Resin and Machine Parameters 154
      6.1.1 Resin-Dependent Parameters 154
      6.1.2 Mold Shrinkage and Processing Temperature 155
      6.1.3 Drying Temperatures and Times 157
   6.2 Melting in Injection Molding Screws 157
      6.2.1 Model 158
      6.2.2 Results of Simulation 161
      6.2.3 Screw Dimensions 163
   6.3 Injection Mold 163
      6.3.1 Runner Systems 163
      6.3.2 Mold Filling 166
   6.4 Flow Characteristics of Injection Molding Resins 168
      6.4.1 Model 169
      6.4.2 Melt Viscosity and Power Law Exponent 170
      6.4.3 Experimental Results and Discussion 171
   6.5 Cooling of Melt in the Mold 174
      6.5.1 Thermal Design of the Mold 174

6.6   Mechanical Design of the Mold 182

6.7   Rheological Design of the Mold 184

References 187

Summary 189

Appendix: List of Programs with Brief Descriptions 191

Index 193

Biography 195

Preface

The plastics engineer working on the shop floor of an industry manufacturing blown film or blow-molded articles or injection-molded parts, to quote a few processes, needs often quick answers to questions such as why the extruder output is low or whether he can expect better quality product by changing the resin or how he can estimate the pressure drop along the runner or gate of an injection mold. Applying the state of the art numerical analysis to address these issues is time-consuming and costly requiring trained personnel. Indeed, as experience shows, most of these issues can be addressed quickly by applying proven, practical calculation procedures which can be handled by pocket calculators and hence can be performed right on the site where the machines are running.

The underlying principles of design formulas for plastics engineers with examples have been treated in detail in the earlier works of Natti Rao.

Bridging the gap between theory and practice this book presents analytical methods based on these formulas which enable the plastics engineer to solve day to day problems related to machine design and process optimization quickly. Basically, the diagnostical approach used here lies in examining whether the machine design is suited to accomplish the desired process parameters.

Starting from solids transport, melting and moving on to shaping the melt in the die to create the product, this work shows the benefits of using simple analytical procedures for troubleshooting machinery and processes by verifying machine design first and then, if necessary, optimizing it to meet the process requirements. Illustrative examples chosen from rheology, heat transfer in plastics processing, extrusion screw and die design, blown film, extrusion coating and injection molding, to mention a few areas, clarify this approach in detail.

Case studies related to melt fracture, homogeneity of the melt, effect of extrusion screw geometry on the quality of the melt, classifying injection molding resins on the basis of their flow length and calculating runner and gate pressure drop are only a few of the topics among many, which have been treated in detail. In addition, parametric studies of blown film, pipe extrusion, extrusion coating, sheet extrusion, thermoforming, and injection molding are presented, so that the user is acquainted with the process targets of the calculations.

The same set of equations can be used to attain different targets whether they deal with extrusion die design or injection molds. Practical calculations illustrate how a variety of goals can be reached by applying the given formulas along with the relevant examples.

In order to facilitate easy use, the formulas have been repeated in some calculations, so that the reader need not refer back to these formulas given elsewhere in the book.

This book is an example-based practical tool not only for estimating the effect of design and process parameters on the product quality but also for troubleshooting practical problems encountered in various fields of polymer processing. It is intended for beginners as well as for practicing engineers, students and teachers in the field of plastics engineering and also for scientists from other areas who deal with polymer engineering in their professions.

The Appendix contains a list of easily applicable computer programs for designing extrusion screws and dies. These can be obtained by contacting the author via email (raonatti@t-online.de).

We are indebted to Dr. Guenter Schumacher of the European Joint Research Center, Ispra, Italy for his generous help in preparing the manuscript. Thanks are also due to Dr. Benjamin Dietrich of KIT, Karlsruhe, Germany for his cooperation in writing the manuscript. The fruitful discussions with Dr. Ranganath Shastri of CIATEQ Unidad Edomex, Mexico are thankfully acknowledged.

The permission of BASF SE for using the main library at Ludwigshafen Rhein, Germany is thankfully acknowledged by Natti S. Rao.

Natti S. Rao, Ph. D.
Nick R. Schott, Ph. D.

1

Rheological Properties of Molten Polymers

The basic principle of making parts from polymeric materials lies in creating a melt from the solid material and forcing the melt into a die, the shape of which corresponds to the shape of the part. Thus, as Fig. 1.1 indicates, melt flow and heat transfer play an important role in the operations of polymer processing.

Plastics solids

Plastication

Melt

Shaping

Cooling

Part removal   **FIGURE 1.1** Principle of manufacturing of plastics parts

■ 1.1 Polymer Melt Flow

Macromolecular fluids, such as thermoplastic melts, exhibit significant non-Newtonian behavior. This is noticed in the marked decrease of melt viscosity when the melt is subjected to shear or tension as shown in Fig. 1.2 and 1.5. The flow of melt in the channels of dies and polymer processing machinery is mainly shear flow. Therefore, knowledge of the laws of shear flow is necessary for designing machines and dies for polymer processing. For practical applications the following summary of the relationships was found to be useful.

1.1.1 Apparent Shear Rate

The apparent shear rate for a melt flowing through a capillary is defined as

$$\dot{\gamma}_a = \frac{4\,\dot{Q}}{\pi\,R^3} \tag{1.1}$$

where $\dot{Q}$ is the volumetric flow rate per second and R is the radius of the capillary. This relationship is for steady state, incompressible flow with no entrance or exit effects, no wall slip and symmetry about the center line.

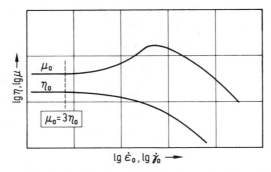

FIGURE 1.2 Tensile viscosity and shear viscosity of a polymer melt as a function of strain rate [4]

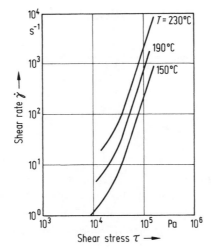

FIGURE 1.3 Flow curves of a LDPE [11]

1.1.2 Apparent Viscosity

The apparent viscosity η_a is defined as

$$\eta_a = \frac{\tau}{\dot{\gamma}_a} \tag{1.2}$$

and is shown in Fig. 1.4 as a function of shear rate and temperature for a LDPE.

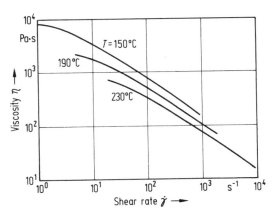

FIGURE 1.4 Viscosity functions of a LDPE [11]

1.1.3 Power Law of Ostwald and De Waele

The power law of Ostwald [7] and De Waele [8] is easy to use, hence widely employed in design work [2]. This relationship can be expressed as

$$\dot{\gamma}_a = K\,\tau^n \tag{1.3}$$

or

$$\dot{\gamma}_a = K\left|\tau^{n-}\right|\tau \tag{1.4}$$

where K denotes a factor of proportionality and n the power law exponent. Another form of power law often used is

$$\tau_a = K_R\,\dot{\gamma}_a^{n_R} \tag{1.5}$$

or

$$\tau_a = K_R\left|\dot{\gamma}_a^{n_R-1}\right|\dot{\gamma}_a \tag{1.6}$$

In this case, n_R is the reciprocal of n and $K_R = K^{-n_R}$ (in the US n_R is used instead of n).

From Eq. (1.3) the exponent n can be expressed as

$$n = \frac{d \lg \dot{\gamma}_a}{d \lg \tau} \tag{1.7}$$

(lg means logarithm to the base of 10 throughout this book)

As shown in Fig. 1.5, in a double log-plot the exponent n represents the local gradient of the curve $\dot{\gamma}_a$ vs. τ.

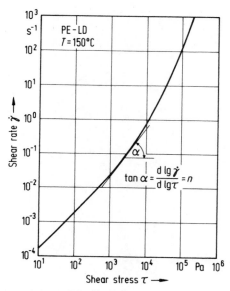

FIGURE 1.5 Determination of the power law exponent *n* in the Eq. (1.9)

1.1.4 Viscosity Formula of Klein

Due to the ease of its application the Klein model, among other rheological models, is best suited for practical use.

The regression equation of Klein et al. [10] is given by

$$\lg \eta_a = a_0 + a_1 \ln \dot{\gamma}_a + a_{11} \left(\ln \dot{\gamma}_a \right)^2 + a_2 T + a_{22} T^2 + a_{12} T \ln \dot{\gamma}_a \tag{1.8}$$

T = Temperature of the melt (°F)

η_a = Viscosity (lb$_f$·s/in^2)

(ln means natural logarithm throughout the text)

The resin-dependent constants a_0 to a_{22} can be determined with the help of the computer program VISRHEO mentioned in the Appendix.

Calculated Example

The following constants are valid for a particular type of LDPE. What is the viscosity η_a at $\dot{\gamma}_a = 500\ s^{-1}$ and $T = 200\ °C$?

$$a_0 \quad = 3.388$$
$$a_1 \quad = -6.351 \cdot 10^{-1}$$
$$a_{11} = -1.815 \cdot 10^{-2}$$
$$a_2 \quad = -5.975 \cdot 10^{-3}$$
$$a_{22} = -2.51 \cdot 10^{-6}$$
$$a_{12} = 5.187 \cdot 10^{-4}$$

$$T\ (°F) = 1.8 \cdot T\ (°C) + 32 = 1.8 \cdot 200 + 32 = 392$$

With the above constants and Eq. (1.8) one gets

$$\eta_a = 0.066\ lb_f \cdot s/in^2$$

and in SI-units

$$\eta_a = 6857 \cdot 0.066 = 449.8\ Pa \cdot s$$

The expression for the power law exponent n can be derived from Eq. (1.8). The exponent n is given by

$$\frac{1}{n} = 1 + a_1 + 2\,a_{11}\ \ln \dot{\gamma}_a + a_{12} \cdot T \tag{1.9}$$

Putting the constants $a_1, \dots, a_{12}$ into this equation one obtains

$$n = 2.919$$

1.1.5 Resin Characterization by Power Law Exponent

It is seen from Fig. 1.6 that n depends less and less on the shear rate with increasing shear rate, so that using a single value of n for a given resin leads to a good estimate of the parameter desired. Plots of n for LLDPE and PET are given in Fig. 1.7 and Fig. 1.8.

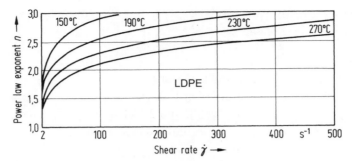

FIGURE 1.6 Power law exponent n as a function of shear rate $\dot{\gamma}$ and melt temperature T [1]

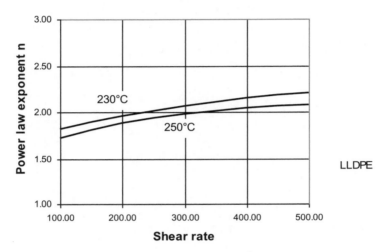

FIGURE 1.7 Power law exponent n as a function of shear rate $\dot{\gamma}$ (s^{-1}) and temperature for LLDPE

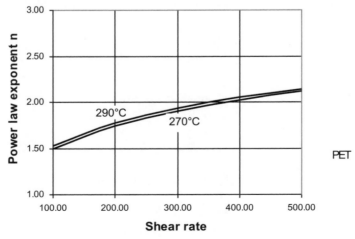

FIGURE 1.8 Power law exponent n as a function of shear rate $\dot{\gamma}$ (s^{-1}) and temperature for PET

■ 1.2 Melt Flow Index

The *Melt Flow Index* (MFI), which is also known as the *Melt Flow Rate* (MFR), indicates the flowability of a constant polymer melt, and is measured by forcing the melt through a capillary under a dead load at constant temperature (Fig. 1.9). The MFI value is the mass of melt flowing in a certain time. A MFR or MFI of 2 at 200 °C and 2.16 kg means, for example, that the melt at 200 °C flows at a rate of 2 g in ten minutes under a dead load of 2.16 kg.

In the case of melt volume rate, which is also known as *Melt Volume Index* (MVI), the volume flow rate of the melt instead of the mass flow rate is set as the basis. The unit here is ml/10 min.

Ranges of melt indices for common processing operations are given in Table 1.1 [6].

TABLE 1.1 Ranges of MFI Values (ASTM D1238) for Common Processes [6]

Process	MFI range
Injection molding	5–100
Rotational molding	5–20
Film extrusion	0.5–6
Blow molding	0.1–1
Profile extrusion	0.1–1

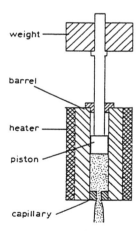

FIGURE 1.9 Melt flow tester [3]

■ 1.3 Relationship between Flow Rate and Pressure Drop

To illustrate the behavior of polymer melt flow, the flow rate is plotted against die pressure for water and LDPE in Fig. 1.10. The diameter and length of the nozzle used are 1 mm and 30 mm, respectively. Fig. 1.10 shows that the laminar flow rate of water increases linearly with the pressure, whereas in the case of LDPE the increase is exponential. To put it in numbers, increasing pressure of water tenfold would bring forth a tenfold flow rate. A tenfold pressure of LDPE, however, leads to about five hundredfold flow rate. The flow of the melt is indeed viscous, but pressure changes are accompanied by much larger changes in output. The relationship between volume flow rate and pressure drop of the melt in a die can be expressed in the general form [14]

$$\dot{Q} = K \cdot G^n \cdot \Delta p^n \qquad (1.10)$$

where

Q = volumetric flow rate

G = die constant

Δp = pressure difference

K = factor of proportionality

n = power law exponent

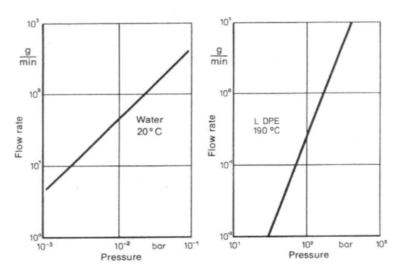

FIGURE 1.10 Flow curves of water in laminar flow and LDPE [2]

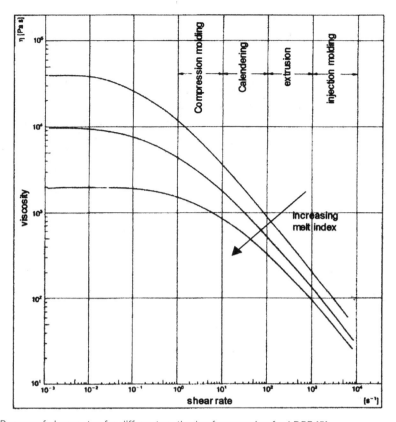

FIGURE 1.11 Ranges of shear rates for different methods of processing for LDPE [5]

The approximate ranges of shear rates for different methods of polymer processing are shown in Fig. 1.11 [5].

Choosing a value of $n = 2.7$ for LDPE, Eq. (1.10) can be written for a given die and processing conditions

$$\dot{Q} \sim \Delta p^{2.7}.\tag{1.11}$$

Increasing the pressure by tenfold would increase the volume flow rate by about five hundred-fold ($10^{2.7} = 501.2$).

■ 1.4 Shear Rates for Extrusion Dies

The formulas for calculating the shear rates in extrusion dies are presented in Table 1.2 [1, 13] and the channel shapes in Fig. 1.12 [9]. Other channel shapes can be taken into account by the formula developed by Schenkel [12].

TABLE 1.2 Shear Rates and Geometry Constants for Some Die Channel Shapes

Channel shape	Shear rate $\dot\gamma$ [s^{-1}]	Geometry constant G
Circle	$\dfrac{4\dot{Q}}{\pi\,R^3}$	$\left(\dfrac{\pi}{4}\right)^{\frac{1}{n}}\cdot\dfrac{R^{\frac{3}{n}+1}}{2\,L}$
Slit	$\dfrac{6\dot{Q}}{W\cdot H^2}$	$\left(\dfrac{W}{6}\right)^{\frac{1}{n}}\cdot\dfrac{H^{\frac{2}{n}+1}}{2\,L}$
Annulus	$\dfrac{6\dot{Q}}{\pi\left(R_o+R_i\right)\left(R_o-R_i\right)^2}$	$\left(\dfrac{\pi}{6}\right)^{\frac{1}{n}}\cdot\dfrac{\left(R_o+R_i\right)^{\frac{1}{n}}\cdot\left(R_o-R_i\right)^{\frac{2}{n}+1}}{2\,L}$
Triangle	$\dfrac{10}{3}\cdot\dfrac{\dot{Q}}{d^3}$	$\dfrac{1}{\sqrt{3}}\cdot\left(\dfrac{3}{10}\right)^{\frac{1}{n}}\cdot\dfrac{d^{\frac{3}{n}+1}}{2\,L}$
Square	$\dfrac{3}{0.42}\cdot\dfrac{\dot{Q}}{a^3}$	$\dfrac{1}{2}\left(\dfrac{0.42}{3}\right)^{\frac{1}{n}}\cdot\dfrac{a^{\frac{3}{n}+1}}{2\,L}$

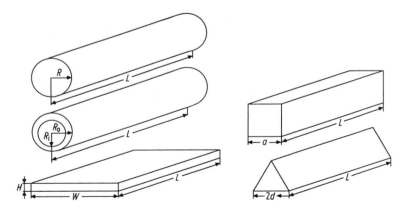

FIGURE 1.12 Common shapes of flow channels in extrusion dies [9]

The following examples illustrate the use of the formulas given above:

Calculated Examples

Example 1

What is the shear rate of a LDPE melt at 200 °C flowing through a round channel of 25 mm diameter at a mass flow rate of $m = 36$ kg/h with a melt density of $\rho_n = 0.7$ g/cm^3?

Solution

Volume shear rate

$$\dot{Q} = \frac{\dot{m}}{\rho_n} = 1.429 \cdot 10^{-5} \text{ m}^3 /\text{s}$$

Shear rate

$$\gamma = \frac{4 \dot{Q}}{\pi R^3} = 9.361 \, \text{s}^{-1}$$

Example 2

Melt flow through an annulus of a blown film die with an outside radius $R_o = 40$ mm and an inside radius $R_i = 39$ mm. The resin is LDPE with the same *viscosity* as above. Mass flow rate and the melt temperature remain the same.

Solution

Volume shear rate

$$\dot{Q} = \frac{\dot{m}}{\rho_n} = 1.429 \cdot 10^{-5} \text{ m}^3 /\text{s}$$

Shear rate

$$\dot{\gamma} = \frac{6 \dot{Q}}{\pi \left(R_o + R_i \right) \left(R_o - R_i \right)^2} = 345.47 \text{ s}^{-1}$$

Example 3

For the same conditions as above it is required to calculate the shear rate when the melt flows through a slit of width $W = 75$ mm and height $H = 1$ mm of an extrusion coating die.

Solution

Shear rate

$$\dot{\gamma} = \frac{6\,\dot{Q}}{W\,H^2} = 1143.2\ \text{s}^{-1}$$

The calculated shear rates in different dies enable one to obtain the shear viscosity from the plot η vs. γ, which then can be used in a comparative study of the resin behavior.

Example 4

Calculation of the pressure drop Δp for the conditions given in Example 2. Using the Klein model [6]

$$\ln \eta = a_0 + a_1 \ln \dot{\gamma} + a_{11}\ \ln \dot{\gamma}^2 + a_2\ T + a_{22}\ T^2 + a_{12}\ T \ln \dot{\gamma}$$

with viscosity coefficients [6]

$$a_0 = 3.388, \quad a_1 = -0.635, \quad a_{11} = -0.01815,$$
$$a_2 = -0.005975, \quad a_{22} = -0.0000025, \quad a_{12} = 0.0005187$$

η at $\dot{\gamma} = 9.316\ \text{s}^{-1}$ is found to be $\eta = 4624.5$ Pa s with $T = 392\ °\text{F}$. The power law exponent n follows from Eq. (1.9)

$$n = 2.052$$

Shear stress

$$\tau = \eta \cdot \dot{\gamma} = 43077\ \text{N/m}^2$$

Factor of proportionality K in Eq. (1.10) from Eq. (1.3)

$$K = 28824 \cdot 10^{-9}$$

Die constant from Table 1.2

$$G_{\text{circle}} = \left(\frac{\pi}{4}\right)^{\frac{1}{2.052}} \frac{0.0125^{\frac{3}{2.052}+1}}{2 \cdot 0.1} = 9.17 \cdot 10^{-5}$$

Finally Δp from Eq. (1.10)

$$\Delta p = \frac{\left(1.429 \cdot 10^{-5}\right)^{\frac{1}{2.052}}}{\left(2.8824 \cdot 10^{-9}\right)^{\frac{1}{2.052}} \cdot 9.17 \cdot 10^{-5}} = 6.845\ \text{bar}$$

References

[1]   Rao, N. S.: Design Formulas for Plastics Engineers, Hanser, Munich, 1991
[2]   Rao, N. S.: Designing Machines and Dies, Polymer Processing, Hanser, Munich 1981
[3]   Rauwendal, C.: Polymer Extrusion 4th ed., Hanser, Munich 2001
[4]   Laun, H. M.: *Progr. Colloid & Polymer Sai*, **75**, 111 (1987)
[5]   N. N.: BASF Brochure: Kunststoff-Physik im Gespräch, 1977
[6]   Rosato, D. V.; Rosato, D. V.: Plastics Processing Data Handbook, Van Nostrand Reinhold, New York 1990
[7]   Ostwald, W.: *Kolloid-Z.*, **36**, 99 (1925)
[8]   De Waele, A.: *Oil and Color Chem. Assoc. J.*, **6**, 33 (1923)
[8]   Bernhardt, E. C.: Processing of Thermoplastic Materials, Reinbold, New York (1963)
[9]   Klein, I.; Marshall, D. I.; Friehe, C. A.: *Soc Plastic Engrs. J.*, **21**, 1299 (1965)
[10] Muenstedt,H.: Kunststoffe, 68 (1978) p. 92
[11] Schenkel, G.: Kunststoff-Extrudiertechnik, Hanser, Munich 1963
[12] Ramsteiner, F.: Kunststoffe 61 (1971) p. 943
[13] Procter, B.: SPE J. 28 (1972) p. 34

2 Thermal Properties of Solid and Molten Polymers

In addition to the mechanical and melt flow properties, thermodynamic data of polymers are necessary for optimizing various heating and cooling processes that occur in plastics processing operations.

In design work the thermal properties are often required as functions of temperature and pressure. As the measured data cannot always be predicted by physical relationships accurately enough, regression equations are used to fit the data for use in design calculations.

■ 2.1 Specific Volume

The volume-temperature relationship as a function of pressure is shown for a semi-crystalline PP in Fig. 2.1 [1] and for an amorphous PS in Fig. 2.2 [1]. The p-v-T diagrams are needed in many applications, for example to estimate the shrinkage of plastics parts in injection molding [19]. Data on p-v-T relationships for a number of polymers are presented in reference [8].

According to the Spencer-Gilmore equation, which is similar to the Van-der-Waal equation of state for real gases, the relationship between pressure p, specific volume v, and temperature T of a polymer can be written as

$$(v - b^\star)(p + p^\star) = \frac{R\,T}{W} \tag{2.1}$$

In this equation, $b^\star$ is the specific individual volume of the macromolecule, $p^\star$ the cohesion pressure, W the molecular weight of the monomer, and R the universal gas constant [9].

The values $p^\star$ and $b^\star$ can be determined from p-V-T diagrams by means of regression analysis. Spencer and Gilmore and other workers evaluated these constants from measurements for the polymers listed in Table 2.1 [9, 18].

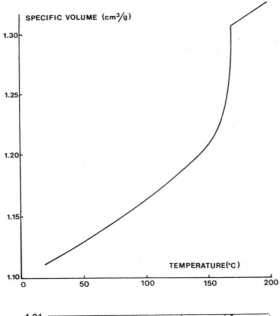

FIGURE 2.1 Specific volume vs. temperature for a semi-crystalline polymer (PP) [1]

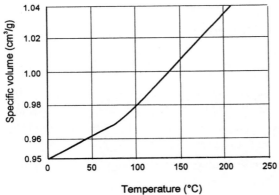

FIGURE 2.2 Specific volume vs. temperature for an amorphous polymer (PS) [1]

TABLE 2.1 Constants for the Equation of State [9]

Material	W g/mol	p^* atm	b^* cm³/g
LDPE	28.1	3240	0.875
PP	41.0	1600	0.620
PS	104	1840	0.822
PC	56.1	3135	0.669
PA 610	111	10768	0.9064
PMMA	100	1840	0.822
PET	37.0	4275	0.574
PBT	113.2	2239	0.712

Calculated Example

The following values are given for a PE-LD:

W = 28.1 g/mol

b^* = 0.875 cm³/g

p^* = 3240 atm

Calculate the specific volume at

T = 190 °C and p = 1 bar

Using Eq. (2.1) and the conversion factors to obtain the volume v in cm³/g, we obtain

$$v = \frac{10 \cdot 8.314 \cdot (273 + 190)}{28.1 \cdot 3240.99 \cdot 1.013} + 0.875 = 1.292 \text{ cm}^3/\text{g}$$

The density ρ is the reciprocal value of specific volume so that

$$\rho = \frac{1}{v}$$

(2. 2)

The p-v-T data can also be fitted by a polynomial of the form

$$v = A(0)_v + A(1)_v \cdot p + (2)_v \cdot T + A(3)_v \cdot T \cdot p$$

(2.3)

if measured data is available (Fig. 2.3) [10, 11, 17]. The empirical coefficients $A(0)_v \ldots A(3)_v$ can be determined by means of the computer program given in [11]. With the modified two-domain Tait equation [16] a very accurate fit can be obtained both for the solid and melt regions.

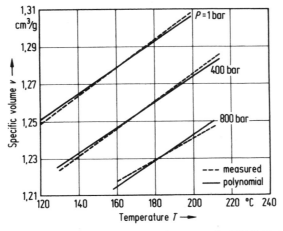

FIGURE 2.3 Specific volume as a function of temperature and pressure for PE-LD [2, 13]

■ 2.2 Specific Heat

The specific heat c_p is defined as (US uses capital letters instead of h)

$$c_p = \left(\frac{\partial h}{\partial T}\right)_p \qquad (2.4)$$

where

h = enthalpy

T = Temperature

The specific heat c_p represents the amount of heat that is supplied to a system in a reversible process at a constant pressure in order to increase the temperature of the substance by dT. The specific heat at constant volume c_v is given by (US uses capital letters instead of u)

$$c_v = \left(\frac{\partial u}{\partial T}\right)_v \qquad (2.5)$$

where

u = internal energy

T = Temperature

In the case of c_v the supply of heat to the system occurs at constant volume.

c_p and c_v are related to each other through the Spencer-Gilmore equation, Eq. (2.1)

$$c_v = c_p - \frac{R}{W} \qquad (2.6)$$

The numerical values of c_p and c_v differ by roughly 10%, so that for approximate calculations c_v can be made equal to c_p. Plots of c_p as a function of temperature are shown in Fig. 2.4 for amorphous, semi-crystalline, and crystalline polymers.

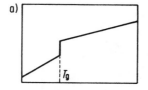  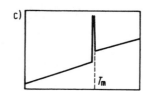

FIGURE 2.4 Specific heat as a function of temperature for amorphous (a), semi-crystalline (b), and crystalline polymers (c) [14]

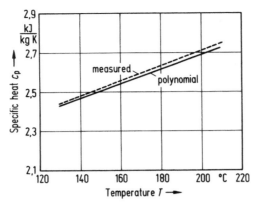

FIGURE 2.5 Comparison between measured values of c_p [13] and polynomial for LDPE [2]

As shown in Fig. 2.5, measured values can be fitted by a polynomial of the type [11]

$$c_p(T) = A(0)\,c_p + A(1)\,c_p \cdot T + A(2)\,c_p \cdot T^2 \tag{2.7}$$

■ 2.3 Thermal Expansion Coefficient

The expansion coefficient α_v at constant pressure is given by [14]

$$\alpha_v = \frac{1}{v}\left(\frac{\partial v}{\partial T}\right)_p \tag{2.8}$$

The isothermal compression coefficient γ_k is defined as [14]

$$\gamma_k = -\frac{1}{v}\left(\frac{\partial v}{\partial p}\right)_T \tag{2.9}$$

α_v and γ_k are related to each other by the expression [14]

$$c_p = c_v + \frac{T \cdot v \cdot \alpha_v^2}{\gamma_k} \tag{2.10}$$

The linear expansion coefficient α_{lin} is approximately

$$\alpha_{lin} = \frac{1}{3}\alpha_v \tag{2.11}$$

Data on Coefficients of Thermal Expansion

Table 2.2 shows the linear expansion coefficients of some polymers at 20 °C. The linear expansion coefficient of mild steel lies around $11 \cdot 10^{-6}$ $[K^{-1}]$ and that of aluminum about $25 \cdot 10^{-6}$ $[K^{-1}]$. As can be seen from Table 2.2, plastics expand about 3 to 20 times more than metals. Factors affecting thermal expansion are crystallinity, cross-linking, and fillers [1].

TABLE 2.2 Coefficients of Linear Thermal Expansion [1, 5]

Polymer	Coefficient of linear expansion at 20 °C a_{lin} 10^6 K^{-1}
PE-LD	250
PE-HD	200
PP	50
PVC-U	75
PVC-P	180
PS	70
ABS	90
PMMA	70
POM	100
PSU	50
PC	65
PET	65
PBT	70
PA 6	80
PA 66	80
PTFE	100
TPU	150

■ 2.4 Enthalpy

Equation (2.4) leads to

$$dh = c_p \cdot dT \tag{2.12}$$

As shown in Fig. 2.6, the measured data for $h = h(T)$ [13] for a polymer melt can be fitted by the polynomial

$$h(T) = A(0)_h + A(1)_h \cdot T + A(2)_h \cdot T^2 \tag{2.13}$$

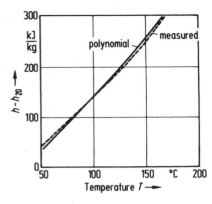

FIGURE 2.6 Comparison between measured values for h [13] and polynomial for PA 6 [11]

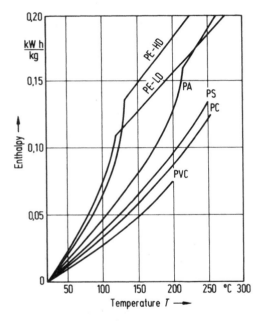

FIGURE 2.7 Specific enthalpy as a function of temperature [14]

The specific enthalpy defined as the total energy supplied to the polymer divided by the throughput of the polymer is a useful parameter for designing extrusion and injection molding equipment such as screws. It provides the theoretical amount of energy required to bring the solid polymer to the process temperature. Values of this parameter for different polymers are given in Fig. 2.7 [14].

If, for example, the throughput of an extruder is 100 kg/h of polyamide (PA) and the processing temperature is 260 °C, the theoretical power requirement would be 20 kW. This can be assumed to be a safe design value for the motor horse power, although theoretically it includes the power supplied to the polymer by the heater bands of the extruder as well.

■ 2.5 Thermal Conductivity

The thermal conductivity λ (US symbol is k) is defined as

$$\lambda = \frac{Q \cdot l}{t \cdot A \cdot (T_1 - T_2)} \tag{2.14}$$

where

Q = heat flow through the surface of area A in a period of time t

$(T_1 - T_2)$ = temperature difference over the length l.

Analogous to the specific heat c_p and enthalpy h, the thermal conductivity can be expressed as [2]

$$\lambda(T) = A(0)_\lambda + A(1)_\lambda \cdot T + A(2)_\lambda \cdot T^2 \tag{2.15}$$

as shown in Fig. 2.8.

The thermal conductivity increases only slightly with pressure. A pressure increase from 1 bar to 250 bar leads only to an increase in thermal conductivity of less than 5% of its value at 1 bar.

As in the case of other thermal properties, thermal conductivity is, in addition to its dependence on temperature, strongly influenced by the crystallinity and orientation and by the amount and type of filler in the polymer [1]. Foamed plastics have, for example, thermal conductivities at least an order of magnitude lower than those of solid polymers [1].

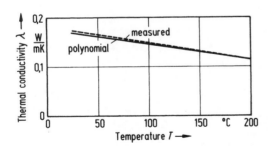

FIGURE 2.8 Comparison between measured values of λ [13] and polynomial for PP [2]

■ 2.6 Thermal Diffusivity

Thermal diffusivity a (in the US α is used instead of a) is defined as the ratio of thermal conductivity to heat capacity per unit volume [1]

$$a = \frac{\lambda}{\rho \cdot c_\mathrm{p}} \tag{2.16}$$

and is of importance in dealing with transient heat transfer phenomena, such as cooling of melt in an injection mold [2]. Although for approximate calculations average values of thermal diffusivity can be used, more accurate computations require functions of λ, ρ, and c_p against temperature for the solid as well as melt regions of the polymer. Thermal diffusivities of some polymers at 20 °C are listed in Table 2.3 [14, 15].

Exhaustive measured data of the quantities c_p, h, λ and pvT diagrams of polymers are given in the VDMA-Handbook [8]. Approximate values of thermal properties of use to plastics engineers are summarized in Table 2.4 [14, 15].

Experimental techniques of measuring enthalpy, specific heat, melting point, and glass transition temperature by differential thermal analysis (DTA) or differential scanning calorimetry (DSC) are described in detail in [12]. Methods of determining thermal conductivity, pvT values, and other thermal properties of plastics are also treated in [12].

TABLE 2.3 Thermal Diffusivities of Polymers at 20 °C

Polymer	Thermal diffusivity at 20 °C a 10^6 m^2/s
PE-LD	0.12
PE-HD	0.22
PP	0.14
PVC-U	0.12
PVC-P	0.14
PS	0.12
PMMA	0.12
POM	0.16
ABS	0.15
PC	0.13
PBT	0.12
PA 6	0.14
PA 66	0.12
PET	0.11

TABLE 2.4 Approximate values for thermal properties of some polymers [14]

Polymer	Thermal conduc-tivity λ (20 °C) W/m·K	Specific heat c_p (20 °C) kJ/kg K	Density ρ (20 °C) g/cm^3	Glass transition temperature T_g °C	Melting point range T_m °C
PS	0.12	1.20	1.06	101	–
PVC	0.16	1.10	1.40	80	–
PMMA	0.20	1.45	1.18	105	–
SAN	0.12	1.40	1.08	115	–
POM	0.25	1.46	1.42	–73	ca. 175
ABS	0.15	1.40	1.02	115	–
PC	0.23	1.17	1.20	150	–
PE-LD	0.32	2.30	0.92	–120/–90	ca. 110
PE-LLD	0.40	2.30	0.92	–120/–90	ca. 125
PE-HD	0.49	2.25	0.95	–120/–90	ca. 130
PP	0.15	2.40	0.91	–10	160/170
PA6	0.36	1.70	1.13	50	215/225
PA 66	0.37	1.80	1.14	55	250/260
PET	0.29	1.55	1.35	70	250/260
PBT	0.21	1.25	1.35	45	ca. 220

■ 2.7 Coefficient of Heat Penetration

The coefficient of heat penetration is used in calculating the contact temperature that results when two bodies of different temperatures are brought into contact with each other [2, 16].

As shown in Table 2.5, the coefficients of heat penetration of metals are much higher than those of polymer melts. Owing to this, the contact temperature of the wall of an injection mold at the time of injection lies in the vicinity of the mold wall temperature before injection.

TABLE 2.5 Coefficients of Heat Penetration of Metals and Plastics [17]

Material	Coefficient of heat penetration b $W \cdot s^{0.5} \cdot m^{-2} \cdot K^{-1}$
Beryllium copper (BeCu25)	$17.2 \cdot 10^3$
Unalloyed steel (C45W3)	$13.8 \cdot 10^3$
Chromium steel (X40Cr13)	$11.7 \cdot 10^3$
Polyethylene (PE-HD)	$0.99 \cdot 10^3$
Polystyrene (PS)	$0.57 \cdot 10^3$
Stainless steel	$7.56 \cdot 10^3$
Aluminum	$21.8 \cdot 10^3$

The contact temperature $\theta_{w_{max}}$ of the wall of an injection mold at the time of injection is [17]

$$\theta_{w_{max}} = \frac{b_w\,\theta_{w_{min}} + b_p\,\theta_M}{b_w + b_p}$$ (2.17)

where

b = coefficient of heat penetration = $\sqrt{\lambda\,\rho\,c}$

$\theta_{w_{min}}$ = temperature before injection

θ_M = melt temperature

Indices w and p refer to mold and polymer, respectively.

Calculated Example

The values given in Table 2.5 refer to the following units of the properties:

Thermal conductivity λ: W/(m·K)

Density ρ: kg/m^3

Specific heat c: kJ/(kg·K)

The approximate values for steel are

l = 50 W/(m·K)

r = 7850 kg/m^3

c = 0.485 kJ/(kg·K)

The coefficient of heat penetration is

$$b = \sqrt{\lambda \cdot \rho \cdot c} = \sqrt{50 \cdot 7.85 \cdot 10^3 \cdot 0.485 \cdot 10^3} = 13.8 \cdot 10^3\ \text{W s}^{0.5} \cdot \text{m}^{-2} \cdot \text{K}^{-1}$$

■ 2.8 Heat Deflection Temperature

Heat deflection temperature (HDT) or the deflection temperature under load (DTUL) is a relative measure of a polymer's ability to retain its shape at elevated temperatures for short duration while supporting a load. In amorphous materials the HDT almost coincides with the glass transition temperature T_g. Crystalline polymers may have lower values of HDT but are dimensionally more stable at elevated temperatures [5]. Additives such as fillers have a more significant effect on crystalline polymers than on amorphous polymers. The heat deflection temperatures are listed for some materials in Table 2.6. Owing to the similarity of the measuring principle, Vicat softening point, HDT, and Martens temperature lie often close to each other (Fig. 2.9) [6].

TABLE 2.6 Heat Deflection Temperatures (HDT) According to the Method A of Measurement [3, 6]

Material	HDT (Method A) °C
PE-LD	35
PE-HD	50
PP	45
PVC	72
PS	84
ABS	100
PC	135
POM	140
PA6	77
PA66	130
PMMA	103
PET	80
PBT	65

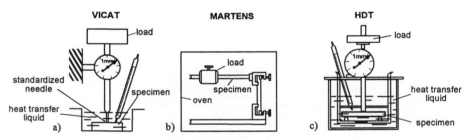

FIGURE 2.9 Principles of measurement of heat distortion of plastics [6]

■ 2.9 Vicat Softening Point

The Vicat softening point represents the temperature at which a small, lightly loaded, heated test probe penetrates a given distance into a test specimen [5].

The Vicat softening point provides an indication of a material's ability to withstand contact with a heated object for a short duration. It is used as a guide value for the demolding temperature in injection molding. Vicat softening points of crystalline polymers have more significance than amorphous polymers, as the latter tend to creep during the test [5]. Both Vicat and HDT values serve as a basis to judge the resistance of a thermoplastic to distortion at elevated temperatures.

Guide values of Vicat softening temperatures of some polymers according to DIN 53460 (Vicat 5 kg) are given in Table 2.7 [6].

TABLE 2.7 Guide Values for Vicat Softening Points [6]

Polymer	Vicat softening point °C
PE-HD	65
PP	90
PVC	92
PS	90
ABS	102
PC	138
POM	165
PA 6	180
PA 66	200
PMMA	85
PET	190
PBT	180

The thermodynamic and viscosity data can be stored in a data bank with the help of the software program, VISRHEO, mentioned in the Appendix and retrieved for design calculations. The Figs. 2.10 and 2.11 show samples of the printouts of this program.

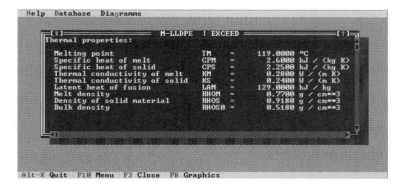

FIGURE 2.10
Thermodynamic data

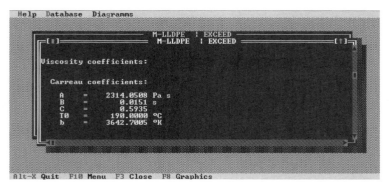

FIGURE 2.11
Viscosity coefficients
in the rheological models

■ References

[1]   Birley, A. W.; Haworth, B.; Bachelor, T.: Physics of Plastics, Hanser, Munich 1991

[2]   Rao, N. S.: Design Formulas for Plastics Engineers, Hanser, Munich, 1991

[3]   Domininghaus, H.: Plastics for Engineers, Hanser, Munich, 1993

[4]   Rigby, R. B.: Polyethersulfone in Engineering Thermoplastics: Properties and Applications. Ed.: James M. Margolis. Marcel Dekker, Basel, 1985

[5]   N. N.: General Electric Plastics Brochure: Engineering Materials Design Guide

[6]   Schmiedel, H.: Handbuch der Kunststoffprüfung (Ed.), Hanser, Munich, 1992

[7]   N. N.: Design Guide, Modern Plastics Encyclopedia, 1978–79

[8]   N. N.: Kenndaten für die Verarbeitung thermoplastischer Kunststoffe, Teil I, Thermodynamik, Hanser, Munich, 1979

[9]   Mckelvey, J. M.: Polymer Processing, John Wiley, New York, 1962

[10] Münstedt, H.: Berechnen von Extrudierwerkzeugen, VDI-Verlag, Düsseldorf, 1978

[11] Rao, N. S.: Designing Machines and Dies Polymer Processing, Hanser, Munich, 1981

[12] N. N.: Brochure Advanced CAE Technology Inc. 1992

[13] N. N.: Proceedings, 9. Kunststofftechnisches Kolloquium, IKV, Aachen, 1978

[14] Rauwendal, C.: Polymer Extrusion 4th ed., Hanser, Munich, 2001

[15] Ogorkiewicz, R. M.: Thermoplastics Properties and Design, John Wiley, New York, 1973

[16] Martin, H.: VDI-Wärmeatlas, VDI-Verlag, Düsseldorf, 1984

[17] Wübken, G.: Berechnen von Spritzgießwerkzeugen, VDI-Verlag, Düsseldorf, 1974

[18] Progelhof, R. C.; Throne, J. L.: Polymer Engineering Principles – Properties, Processes, Tests for Design, Hanser, Munich 1993

[19] Kurfess, W.: *Kunststoffe*, **61** (1971), p. 421

[20] Münstedt, H: *Kunststoffe*, **68**, (1978), p. 92

3 Heat Transfer in Plastics Processing

Heat transfer and flow processes occur in most polymer processing machinery and often determine the production rate. Designing and optimizing machine elements and processes therefore requires the knowledge of the fundamentals of these sciences. The flow behavior of polymer melts has been dealt with in Chapter 1. In the present chapter, the principles of heat transfer of relevance to polymer processing are treated with examples.

■ 3.1 Steady State Conduction

Fourier's law for one dimensional conduction is given by

$$\dot{Q} = -\lambda\, A\, \frac{dt}{dx} \tag{3.1}$$

where

Q = heat flow thermal conductivity

A = area perpendicular to the direction of heat flow

T = temperature

x = distance (Fig. 3.1)

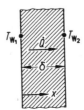

FIGURE 3.1 Plane wall

3.1.1 Plane Wall

Temperature profile (Fig. 3.1) [1]:

$$T(x) = \frac{\left(T_{W_2} - T_{W_1}\right)}{\delta} \cdot x + T_{W_1} \tag{3.2}$$

Heat flow:

$$\dot{Q} = \frac{\lambda}{\delta} \cdot A \cdot \left(T_{W_1} - T_{W_2}\right) \tag{3.3}$$

Analogous to Ohm's law in electric circuit theory, Eq. (3.1) can be written as [2]

$$\dot{Q} = \frac{\Delta T}{R} \tag{3.4}$$

in which

$$R = \frac{\delta}{\lambda \cdot A} \tag{3.5}$$

where

ΔT =   temperature difference

δ =   wall thickness

R =   thermal resistance

Calculated Example

The temperatures of a plastic sheet with a thickness of 30 mm and a thermal conductivity of $\lambda = 0.335$ W/(mK) are according to Fig. 3.1 $T_{W_1} = 100\,°C$ and $T_{W_2} = 40\,°C$. Calculate the heat flow per unit area of the sheet.

Solution:
Substituting the given values in Eq. (3.3) we obtain

$$\frac{\dot{Q}}{A} = \frac{0.335}{(30\,/\,1000)} \cdot (100 - 40) = 670 \text{ W/m}^2$$

3.1.2 Cylinder

Temperature distribution (Fig. 3.2) [1]:

$$T(r) = T_{W_1} + \frac{T_{W_2} - T_{W_1}}{l\left(\dfrac{r_2}{r_1}\right)} \cdot \ln\left(\frac{r}{r_1}\right) \tag{3.6}$$

Heat flow:

$$\dot{Q} = \frac{\lambda}{\delta} \cdot A_m \cdot \left(T_{W_1} - T_{W_2}\right) \tag{3.7}$$

with the log. mean surface area A_m of the cylinder

$$A_m = \frac{A_2 - A_1}{\ln\left(\dfrac{A_2}{A_1}\right)} \tag{3.8}$$

where

$$\delta = r_2 - r_1$$

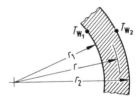

FIGURE 3.2 Steady state heat transfer through a cylindrical wall

3.1.3 Hollow Sphere

Temperature distribution [1]:

$$T(r) = \frac{1}{r} \frac{\left(T_{W_1} - T_{W_2}\right)}{r_2 - r_1} - \frac{\left(T_{W_1} r_1 - T_{W_2} r_2\right)}{r_2 - r_1} \tag{3.9}$$

with the boundary conditions

$$T(r = r_1) = T_{W_1} \quad \text{and} \quad T(r = r_2) = T_{W_2}$$

Heat flow:

$$\dot{Q} = \frac{\lambda}{\delta} \cdot A_m \cdot \left(T_{W_1} - T_{W_2} \right) \tag{3.10}$$

The geometrical mean area A_m of the sphere is

$$A_m = \sqrt{A_1\, A_2} \tag{3.11}$$

The wall thickness δ is

$$\delta = r_2 - r_1. \tag{3.12}$$

3.1.4 Sphere

Heat flow from a sphere in an infinite medium $(r_2 \rightarrow \infty)$ [1]

$$\dot{Q} = 4\,\pi\,r_1\,\lambda \left(T_{W_1} - T_\infty \right) \tag{3.13}$$

where

T_∞ =   temperature at a very large distance.

Figure 3.3 shows the temperature profiles of the one dimensional bodies treated above [1].

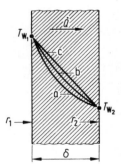

FIGURE 3.3 One dimensional heat transfer; a: sphere, b: cylinder, c: plate

3.1.5 Heat Conduction in Composite Walls

Following the electrical analogy, heat conduction through a multiple layer wall can be treated as a current flowing through resistances connected in series. From this concept we obtain the expression for the heat flow through the composite wall shown by equation 3.17 and Fig. 3.4.

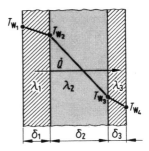

FIGURE 3.4 Heat transfer through a composite wall [1]

$$T_{W_1} - T_{W_2} = \dot{Q}\,\frac{\delta_1}{\lambda_1\,A_1} \tag{3.14}$$

$$T_{W_2} - T_{W_3} = \dot{Q}\,\frac{\delta_2}{\lambda_2\,A_2} \tag{3.15}$$

$$T_{W_3} - T_{W_4} = \dot{Q}\,\frac{\delta_3}{\lambda_3\,A_3} \tag{3.16}$$

Adding Eq. (3.14) to Eq. (3.16) and putting $A_1 = A_2 = A_3 = A$ gives

$$T_{W_1} - T_{W_4} = \dot{Q}\left(\frac{\delta_1}{\lambda_1\,A} + \frac{\delta_2}{\lambda_2\,A} + \frac{\delta_3}{\lambda_3\,A}\right) = \Delta T \tag{3.17}$$

Thus

$$\dot{Q} = \frac{\Delta T}{\left(\dfrac{\delta_1}{\lambda_1\,A} + \dfrac{\delta_2}{\lambda_2\,A} + \dfrac{\delta_3}{\lambda_3\,A}\right)} \tag{3.18}$$

Inserting the conduction resistances

$$R_1 = \frac{\delta_1}{\lambda_1\,A_1} \tag{3.19}$$

$$R_2 = \frac{\delta_2}{\lambda_2\,A_2} \tag{3.20}$$

$$R_3 = \frac{\delta_3}{\lambda_3\,A_3} \tag{3.21}$$

into Eq. (3.18) we get

$$\dot{Q} = \frac{\Delta T}{\left(R_1 + R_2 + R_3\right)} = \frac{\Delta T}{R} \tag{3.22}$$

Calculated Example

A two-layer wall consists of following insulating materials (see Fig. 3.4):

$$\delta_1 = \ \ 16 \text{ mm}, \quad \lambda_1 = 0.048 \text{ W/(m K)}$$
$$\delta_2 = 140 \text{ mm}, \quad \lambda_2 = 0.033 \text{ W/(m K)}$$

The temperatures are $T_{W_1} = 30\ ^\circ\text{C}$, $T_{W_2} = 2\ ^\circ\text{C}$. Calculate the heat loss per unit area of the wall.

$$\Delta T = T_{W_1} T_{W_2} = 30 - 2 = 28\ ^\circ\text{C}$$

Area $A = 1 \text{ m}^2$

$$R_1 = \frac{\delta_1}{\lambda_1 A} = \frac{16 / 1000}{0.048 \cdot 1} = 0.33 \text{ K/W}$$

$$R_2 = \frac{\delta_2}{\lambda_2 A} = \frac{140 / 1000}{0.033 \cdot 1} = 4.24 \text{ K/W}$$

$$\dot{Q} = \frac{\Delta T}{\left(R_1 + R_2\right)} = \frac{28}{\left(4.24 + 0.33\right)} = 6.13 \text{ W}$$

The following example [2] illustrates the calculation of heat transfer through a tube shown in Fig. 3.5.

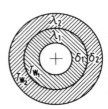

FIGURE 3.5 Heat flow in a multilayered cylinder

Calculated Example

A tube with an outside diameter of 60 mm is insulated with the following materials:

$$\delta_1 = 50 \text{ mm}, \quad \lambda_1 = 0.055 \text{ W/(m K)}$$
$$\delta_2 = 40 \text{ mm}, \quad \lambda_2 = 0.05 \ \text{ W/(m K)}$$

The temperatures are $T_{W_1} = 150 \, °C$ and $T_{W_2} = 30 \, °C$. Calculate the heat loss per unit length of the tube.

Solution:

Resistance R_1:

$$R_1 = \frac{\delta_1}{\lambda_1 \, \overline{A_1}} = \frac{0.05}{0.055 \cdot 2 \, \pi \, \overline{r_1} \cdot L} \, \text{K/W}$$

average radius $\overline{r_1}$:

$$\overline{r_1} = \frac{(80 - 30)}{\ln\left(\dfrac{80}{30}\right)} = 50.97 \text{ mm}$$

$$R_1 = \frac{2.839}{L} \, \text{K/W}$$

Resistance R_2:

$$R_2 = \frac{\delta_2}{\lambda_2 \, \overline{A_2}} = \frac{0.04}{0.05 \cdot 2 \, \pi \, \overline{r_2} \cdot L} \, \text{K/W}$$

average radius $\overline{r_2}$:

$$\overline{r_2} = \frac{(120 - 80)}{\ln\left(\dfrac{120}{80}\right)} = 98.64 \text{ mm}$$

$$R_2 = \frac{1.291}{L} \, \text{K/W}$$

Heat loss per unit length of the tube according to Eq. (3.22):

$$\frac{\dot{Q}}{L} = \frac{150 - 30}{2.839 + 1.291} = 29.1 \, \text{W/m}$$

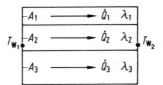

FIGURE 3.6 Heat transfer in composite walls in parallel

In the case of a multi-layer wall, in which the heat flow is divided into parallel flows as shown in Fig. 3.6, the total heat flow is the sum of the individual heat flows. Therefore we have

$$\dot{Q} = \sum_{i=1}^{z} \dot{Q}_i \tag{3.23}$$

$$\dot{Q} = \sum_{i=1}^{z} \left(\lambda_i \, A_i \right) \frac{T_{W_1} - T_{W_2}}{\delta} \tag{3.24}$$

3.1.6 Overall Heat Transfer through Composite Walls

If heat exchange takes place between a fluid and a wall, as shown in Fig. 3.7, we have to consider convection resistance in addition to the conduction resistance, which can be written as

$$R_{c_1} = \frac{1}{\alpha_1 \, A_1} \tag{3.25}$$

where α_i = heat transfer coefficient in the boundary layer near the walls adjacent to the fluids.

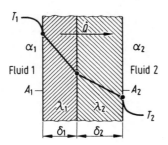

FIGURE 3.7 Conduction and convection through a composite wall

The combination of convection and conduction in stationary walls is called the overall heat transfer and can be expressed as

$$\dot{Q} = k \, A \left(T_1 - T_2 \right) = \frac{1}{R_W} \left(T_1 - T_2 \right) \tag{3.26}$$

where k is denoted as the overall heat transfer coefficient with the corresponding overall resistance R_W

$$R_W = \frac{1}{k\,A} \tag{3.27}$$

Analogous to conduction for the composite wall in Fig. 3.7 the overall resistance R_W can be given by

$$R_W = R_{c_1} + \sum_{i=1}^{z} R_i + R_{c_2} \tag{3.28}$$

or

$$\frac{1}{k\,A} = \frac{1}{\alpha_1\,A_1} + \sum_{i=1}^{z} \frac{\delta_i}{\lambda_i\,A_i} + \frac{1}{\alpha_2\,A_2} \tag{3.29}$$

A simplified form of Eq. (3.29) is

$$k = \frac{1}{\dfrac{1}{\alpha_1} + \displaystyle\sum_{i=1}^{z} \dfrac{\delta_i}{\lambda_i} + \dfrac{1}{\alpha_2}} \tag{3.30}$$

Calculation of the convection heat transfer coefficient is shown in the Section 3.5.

■ 3.2 Unsteady State Conduction

The differential equation for the transient one-dimensional conduction after Fourier is given by

$$\frac{\partial T}{\partial t} = a\,\frac{\partial^2 T}{\partial x^2} \tag{3.31}$$

where

T = temperature

t = time

x = distance

The thermal diffusivity a in this equation is defined as

$$a = \frac{\lambda}{\rho\,c_p} \tag{3.32}$$

where

A = thermal conductivity

c_p = specific heat at constant pressure

p = ensity

For commonly occurring geometrical shapes analytical expressions for transient conduction are given in the following sections.

3.2.1 Temperature Distribution in One-Dimensional Solids

The expression for the heating or cooling of an infinite plate [2] follows from Eq. (3.31) (Fig. 3.8):

$$\frac{T_W - \overline{T_b}}{T_W - T_a} = \frac{8}{\pi^2}\left(e^{-a_1 F_0} + \frac{1}{9} e^{-9 a_1 F_0} + \frac{1}{25} e^{-25 a_1 F_0} + \dots \right) \tag{3.33}$$

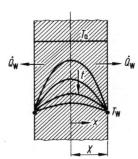

FIGURE 3.8 Unsteady-state conduction in an infinite plate

The Fourier number F_0 is defined by

$$F_0 = \frac{a \cdot t_k}{X^2} \tag{3.34}$$

where

T_W = constant surface temperature of the plate

T_a = initial temperature

$\overline{T_b}$ = average temperature of the plate at time t_T

t_k = heating or cooling time

X = half thickness of the plate $\left[X = \dfrac{s}{2} \right]$

a_1 = $(\pi/2)^2$

a = thermal diffusivity, Eq. (3.32)

The equation for an infinite cylinder with the radius r_m is given by [2]

$$\frac{T_W - \overline{T_b}}{T_W - T_a} = 0.692\,e^{-5.78\,F_0} + 0.131\,e^{-30.5\,F_0} + 0.0534\,e^{-74.9\,F_0} + \dots \qquad (3.35)$$

and for a sphere with the radius r_m

$$\frac{T_W - \overline{T_b}}{T_W - T_a} = 0.608\,e^{-9.87\,F_0} + 0.152\,e^{-39.5\,F_0} + 0.0676\,e^{-88.8\,F_0} + \dots \qquad (3.36)$$

where

$$F_0 = \frac{a \cdot t_k}{r_m^2} \qquad (3.37)$$

In the range $F_0 > 1$ only the first term of these equations is significant. Therefore, we obtain for the heating or cooling time [2]

Plate:

$$t_k = \frac{1}{a} \cdot \left(\frac{s}{\pi}\right)^2 \ln\left[\left(\frac{8}{\pi^2}\right)\left(\frac{T_W - T_a}{T_W - \overline{T_b}}\right)\right] \qquad (3.38)$$

Cylinder:

$$t_k = \frac{r_m^2}{5.78\,a} \ln\left[0.692\left(\frac{T_W - T_a}{T_W - \overline{T_b}}\right)\right] \qquad (3.39)$$

Sphere:

$$t_k = \frac{r_m^2}{9.87\,a} \ln\left[0.608\left(\frac{T_W - T_a}{T_W - \overline{T_b}}\right)\right] \qquad (3.40)$$

The solutions of Eq. (3.32) to Eq. (3.40) are presented in a semi logarithmic plot in Fig. 3.9, in which the temperature ratio $\Theta_{\overline{T_b}} = (T_W - \overline{T_b})/(T_W - T_a)$ is shown as a function of the Fourier number F_0.

Excepting small Fourier numbers, these plots are straight lines approximated by the Eq. (3.38) to Eq. (3.40).

If the time t_k is based on the center line temperature T_b instead of the average temperature $\overline{T_b}$, then we get [3]

$$t_k = \frac{s^2}{\pi^2 \cdot a} \ln\left[\frac{4}{\pi}\left(\frac{T_W - T_a}{T_W - \overline{T_b}}\right)\right] \qquad (3.41)$$

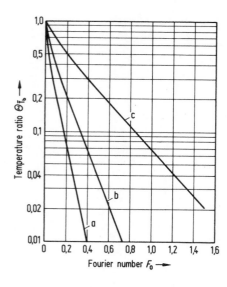

FIGURE 3.9 Average temperature of an infinite slab (c), a long cylinder (b), and a sphere (a) during unsteady heating or cooling

Analogous to Fig. 3.9, the ratio $\Theta_{\overline{T}_b}$ with the center line temperature T_b at time t_k is plotted in Fig. 3.10 over the Fourier number for bodies of different geometries [4].

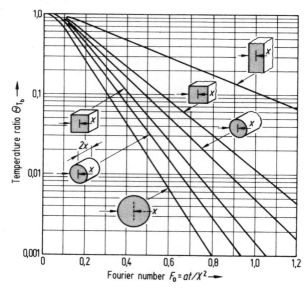

FIGURE 3.10 Axis temperature for multidimensional bodies

The foregoing equations apply to the case, in which the thermal resistance between the body and the surroundings is negligibly small ($\alpha_0 \rightarrow \infty$), for instance in injection molding between the part and the coolant. This means that the Biot number should be very large, $Bi \rightarrow \infty$. The Biot number for a plate is

$$Bi_{plate} = \frac{\alpha_a \cdot X}{\lambda} \qquad (3.42)$$

where

α_a = heat transfer coefficient of the fluid

λ = thermal conductivity of the plastic

As the heat transfer coefficient in practice has a finite value, the temperature ratio Θ_{T_b} based on the center line temperature is given in Fig. 3.11 as a function of the Fourier number with the reciprocal of the Biot number as a parameter [5].

Calculated Examples

Example 1

Cooling of a part in an injection mold for the following conditions:

Resin: LDPE

Thickness of the plate	s	= 12.7 mm
Temperature of the melt	T_a	= 243.3 °C
Mold temperature	T_W	= 21.1 °C
Demolding temperature	T_b	= 76.7 °C
Thermal diffusivity	a	= $1.29 \cdot 10^{-3}$ cm²/s

The cooling time t_k is to be calculated.

Solution:

The temperature ratio Θ_{T_b} :

$$\Theta_{T_b} = \frac{T_b - T_W}{T_W - T_a} = \frac{76.7 - 21.1}{243.3 - 21.1} = 0.25$$

Fourier number F_0 from Fig. 3.10 at $\Theta_{T_b} = 0.25$

$$F_0 = 0.65$$

Cooling time t_k:

$$X = \frac{s}{2} = \frac{12.7}{2} = 6.35 \text{ mm}$$

$$F_0 = \frac{a \cdot t_k}{X^2} = \frac{1.29 \cdot 10^{-3} \cdot t_k}{0.635^2} = 0.65$$

$$t_k = 203 \text{ s}$$

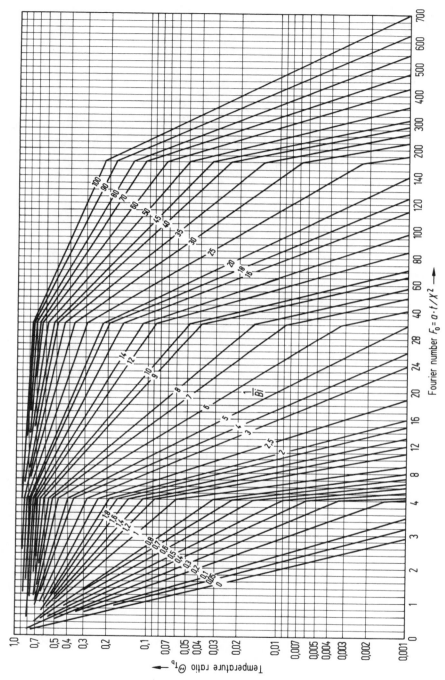

FIGURE 3.11 Midplane temperature for an infinite plate

Example 2

Calculate the cooling time t_k in the Example 1 if the mold is cooled by a coolant having a heat transfer coefficient of $\alpha_a = 2839 \ W/(m^2 \cdot K)$.

Solution:

$$\alpha_a = 2839 \ W/(m^2 \cdot K)$$

$$\lambda_{plastic} = 0.242 \ W/(m \cdot K)$$

The resulting Biot number is

$$Bi = \frac{\alpha_a \cdot X}{\lambda} = \frac{2839 \cdot 6.35}{0.242 \cdot 1000} = 74.49 \qquad \frac{1}{Bi} = 0.01342$$

As can be read from Fig. 3.11, the Fourier number does not differ much from the one in the previous example for $\Theta_{T_b} = 0.25$ and $1/Bi = 0.01342$. The resistance due to convection is therefore negligible and the cooling time remains almost the same. However, the convection resistance has to be taken into account in the case of a film with a thickness of 127 µ cooling in quiescent air, as the following calculation shows: The heat transfer coefficient for this case is approximately

$$\alpha_a = 56.78 \ W/(m^2 \cdot K)$$

The Biot number Bi is

$$Bi = \frac{56.78 \cdot 63.5}{10^6 \cdot 0.242} = 0.0149 \qquad \frac{1}{Bi} = 67.1$$

F_0 from Fig. 3.11

$$F_0 = 95$$

The cooling time is

$$t_k = \frac{X^2 \cdot F_0}{a} = \frac{65.5^2 \cdot 10^3 \cdot 95}{10^8 \cdot 1.29} = 2.97 \ s$$

Example 3

Cooling of an extruded wire: a strand of polyacetal of diameter 3.2 mm is extruded at 190 °C into a water bath at 20 °C. Calculate the length of the water bath to cool the strand from 190 °C to a center line temperature of 140 °C. The following conditions are given:

$$\alpha_a = 1700 \ W/(m^2 \cdot K)$$

$$a_{plastic} = 10^{-7} \ m^2/s$$

$$\lambda_{plastic} = 0.23 \ W/(m \cdot K)$$

Haul-off rate of the wire $V_H = 0.5 \ m/s$

Solution:

The Biot number is

$$Bi = \frac{\alpha_a \cdot X}{\lambda}$$

where R = radius of the wire

$$Bi = \frac{1700 \cdot 1.6}{1000 \cdot 0.23} = 11.13 \qquad \frac{1}{Bi} = 0.0846$$

The temperature ratio Θ_{T_b}

$$\Theta_{T_b} = \frac{T_b - T_W}{T_W - T_a} = \frac{140 - 20}{190 - 20} = 0.706$$

The Fourier number F_0 for $\Theta_{T_b} = 0.706$ and $\frac{1}{Bi} = 0.0846$ from Fig. 3.11 is roughly

$$F_0 \approx 0.16$$

The cooling time t_k follows from

$$\frac{a \cdot t_k}{R^2} = \frac{10^{-7} \cdot t_k}{\left(1.6 \cdot 10^{-3}\right)^2} = \frac{t_k}{2.56 \cdot 10} = 0.16$$

$$t_k = 4.1 \ s$$

The length of the water bath is

$$V_H \cdot t_k = 0.5 \cdot 4.1 = 2.05 \ m$$

3.2.2 Thermal Contact Temperature

If two semi infinite bodies of different initial temperatures Θ_{A_1} and Θ_{A_2} are brought into contact, as indicated in Fig. 3.12, the resulting contact temperature Θ_K is given by [3]

$$\Theta_K = \frac{\Theta_{A_1} + \dfrac{\left(\sqrt{\lambda\,\rho\,c}\right)_2}{\left(\sqrt{\lambda\,\rho\,c}\right)_1}\,\Theta_{A_2}}{1 + \dfrac{\left(\sqrt{\lambda\,\rho\,c}\right)_2}{\left(\sqrt{\lambda\,\rho\,c}\right)_1}} \tag{3.43}$$

where

λ	=	thermal conductivity
ρ	=	density
c	=	specific heat
$\sqrt{\lambda\,\rho\,c}$	=	coefficient of heat penetration

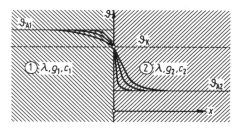

FIGURE 3.12 Temperature distribution in semi-infinite solids in contact

Equation (3.43) also applies to the case of contact of short duration for thick bodies. It follows from this equation that the contact temperature depends on the ratio of the coefficients of heat penetration and lies nearer to the initial temperature of the body that has a higher coefficient of penetration. The ratio of the temperature differences $(\Theta_{A_1} - \Theta_K)$ and $(\Theta_K - \Theta_{A_2})$ are inversely proportional to the coefficient of penetration:

$$\frac{\Theta_{A_1} - \Theta_K}{\Theta_K - \Theta_{A_2}} = \frac{\left(\sqrt{\lambda\,\rho\,c}\right)_2}{\left(\sqrt{\lambda\,\rho\,xc}\right)_1} \tag{3.44}$$

Calculated Example

The contact temperature $\Theta_{w_{min}}$ of the wall of an injection mold at the time of injection is according to Eq. (3.43) [8]

$$\Theta_{w_{max}} = \frac{b_W\, \Theta_{w_{min}} + b_p\, \Theta_M}{b_w + b_p} \qquad (3.45)$$

where

$b \quad = \sqrt{\lambda\, \rho\, c}$

$\Theta_{w_{min}}$ = temperature before injection

Θ_M = melt temperature

Indices w and p refer to mold and polymer, respectively.

As shown in Table 3.1 [8], the coefficients of heat penetration of metals are much higher than those of polymer melts. Hence the contact temperature lies in the vicinity of the mold wall temperature before injection.

TABLE 3.1 Coefficients of Heat Penetration of Mold Material and Resin

Material	Coefficient of heat penetration b ($Ws^{0.5} \cdot m^{-2} \cdot K^{-1}$)
Beryllium copper (BeCu25)	$17.2 \cdot 10^3$
Unalloyed steel (C45W3)	$13.8 \cdot 10^3$
Chromium steel (X40Crl3)	$11.7 \cdot 10^3$
Polyethylene (PE-HD)	$0.99 \cdot 10^3$
Polystyrene (PS)	$0.57 \cdot 10^3$

The values given in the Table 3.1 refer to the following units of the properties:

Thermal conductivity λ: W/(m·K)

Density ρ: kg/m^3

Specific heat c: kJ/(kg·K)

The approximate values for steel are

$\lambda = 50$ W/(m · K)

$\rho = 7850$ kg/m^3

$c = 0.485$ kJ/(kg · K)

The coefficient of heat penetration b

$$b = \sqrt{\lambda \cdot \rho \cdot c} = \sqrt{50 \cdot 7.85 \cdot 10^3 \cdot 0.485 \cdot 10^3} = 13.8 \cdot 10^3 \ Ws^{0.5} \cdot m^{-2} \cdot K^{-1}$$

■ 3.3 Heat Conduction with Dissipation

The power created by the tangential forces in the control volume of the fluid flow is denoted as dissipation [9]. In shear flow, the rate of energy dissipation per unit volume is equal to the product of shear force and shear rate [10]. The power due to dissipation [11] is therefore

$$\dot{E}_d = \tau \cdot \dot{\gamma} \tag{3.46}$$

From the power law we get

$$\dot{E}_d = \left(\frac{1}{K}\right)^{\frac{1}{n}} \cdot \dot{\gamma}^{\frac{1}{n}+1} \tag{3.47}$$

For a Newtonian fluid with $n = 1$ we obtain

$$\dot{E}_d = \eta \cdot \dot{\gamma}^2 \tag{3.48}$$

The applicable differential equation for a melt flow between two plates, where the upper plate is moving with a velocity U_x and the lower plate is stationary [11], is

$$\lambda \frac{\partial^2 T}{\partial y^2} + \eta \left(\frac{\partial u_x}{\partial y}\right)^2 = 0 \tag{3.49}$$

For drag flow the velocity gradient is given by

$$\frac{\partial u}{\partial y} = \frac{U_x}{H} \tag{3.50}$$

Equation (3.49) can now be written as

$$\lambda \frac{\partial^2 T}{\partial y^2} + \eta \left(\frac{U_x}{H}\right)^2 = 0 \tag{3.51}$$

If the temperature of the upper plate is T_1 and that of lower plate T_0, the temperature profile of the melt is obtained by integrating Eq. (3.51). The resulting expression is

$$T = \frac{\eta\, U_x \cdot y}{2\,\lambda\, H}\left(1 - \frac{y}{H}\right) + \frac{y}{H}\left(T_1 - T_0\right) + T_0 \tag{3.52}$$

As shown in Section 4.2.3, this equation can be used to calculate the temperature of the melt film in an extruder.

◼ 3.4 Dimensionless Groups

Dimensionless groups can be used to describe complicated processes which are influenced by a large number of variables with the advantage that the whole process can be analyzed on a sound basis by means of a few dimensionless parameters. Their use in correlating experimental data and in scaling up of equipment is well known. Table 3.2 shows some of the dimensionless groups which are often used in plastics engineering.

TABLE 3.2 Dimensionless Groups

Symbol	Name	Definition
Bi	Biot number	$\alpha_a\, l\, /\, \lambda \infty$
Br	Brinkman number	$\eta\, w^2\, /\, (\lambda\, \Delta T)$
Deb	Deborah number	$t_D\, /\, t_P$
F_0	Fourier number	$a\, t\, /\, l^2$
Gr	Grashof number	$g\, \beta \cdot \Delta T\, l^3\, /\, v^2$
Gz	Graetz number	$l^2\, /\, (a \cdot t_v)$
Le	Lewis number	$a\, /\, \delta$
Na	Nahme number	$\beta_T\, w^2\, \eta\, /\, \lambda$
Nu	Nusselt number	$\alpha\, l\, /\, \lambda$
Pe	Peclet number	$w\, l\, /\, a$
Pr	Prandtl number	$v\, /\, a$
Re	Reynolds number	$\rho\, w\, l\, /\, \eta$
Sh	Sherwood number	$\beta_s\, l\, /\, \delta$
Sc	Schmidt number	$v\, /\, \delta$
Sk	Stokes number	$P \cdot l\, /\, (\eta \cdot w)$

Nomenclature:

a:	Thermal diffusivity	(m^2/s)
g:	Acceleration due to gravity	(m/s^2)
l:	Characteristic length	(m)
p:	Pressure	(N/m^2)
t:	Time	(s)
Indices D, P:	memory and process of polymer respectively	
ΔT:	Temperature difference	(K)
w:	Velocity of flow	(m/s)
α_a:	Outside heat transfer coefficient	$[W/(m^2 \cdot K)]$
β:	Coefficient of volumetric expansion	(K^{-1})
β_T:	Temperature coefficient in the power law of viscosity	(K^{-1})
β_s:	Mass transfer coefficient	(m/s)
δ:	Diffusion coefficient	(m^2/s)
η:	Viscosity	(Ns/m^2)
λ:	Thermal conductivity (index i refers to the inside value)	$[W/(m\ K)]$
v:	Kinematic viscosity	(m^2/s)
t_v:	Residence time	(s)
ρ:	Density	(kg/m^3)

Physical Meaning of Dimensionless Groups

Biot number: Ratio of thermal resistances in series: $(l / \lambda_i) / (1 / \alpha_a)$;
Application: heating or cooling of solids by heat transfer through conduction and convection.

Brinkman number: ratio of heat dissipated $(\eta\ w^2)$ to heat conducted $(\lambda\ \Delta T)$;
Application: polymer melt flow.

Deborah number: ratio of the period of memory of the polymer to the duration of processing [13]. At $Deb > 1$ the process is determined by the elasticity of the material, whereas at $Deb < 1$ the viscous behavior of the polymer influences the process remarkably.

Fourier number: ratio of a characteristic body dimension to an approximate temperature wave penetration depth for a given time [16];
Application: unsteady state heat conduction.

Graetz number: ratio of the time to reach thermal equilibrium perpendicular to the flow direction (l^2 / a) to the residence time (t_v);
Application: heat transfer to fluids in motion.

Grashof number: ratio of the buoyant force $g\beta \cdot \Delta T\, l^3$ to frictional force (v);
Application: heat transfer by free convection.

Lewis number: ratio of thermal diffusivity (a) to the diffusion coefficient (δ);
Application: phenomena with simultaneous heat and mass transfer.

Nahme or Griffith number: ratio of viscous dissipation $(\beta_T\, w^2\, \eta)$ to the heat by conduction (λ) perpendicular to the direction of flow;
Application: heat transfer in melt flow.

Nusselt number: ratio of the total heat transferred $(\alpha \cdot l)$ to the heat by conduction (λ)
Application: convective heat transfer.

Peclet number: ratio of heat transfer by convection $(\rho\, c_p \cdot w \cdot l)$ to the heat by conduction (λ);
Application: heat transfer by forced convection.

Prandtl number: ratio of the kinematic viscosity (v) to thermal diffusivity (a);
Application: convective heat transfer.

Reynolds number: ratio of the inertial force $(\rho\, w\, l)$ to viscous force (η);
Application: the Reynolds number serves as a criterion to judge the type of flow. In pipe flow, when $Re < 2300$, the flow is laminar. The flow is turbulent at approx. $Re > 4000$. Between 2100 and 4000, the flow may be laminar or turbulent, depending on conditions at the entrance of the tube and on the distance from the entrance [2];
Application: fluid flow and heat transfer.

Sherwood number: ratio of the resistance to diffusion $(1/\delta)$ to the resistance to mass transfer $(1/\beta_s)$;
Application: mass transfer problems.

Schmidt number: ratio of kinematic viscosity (v) to the diffusion coefficient (δ);
Application: heat and mass transfer problems.

Stokes number: ratio of pressure forces $(p \cdot l)$ to viscous forces $(\eta \cdot w)$;
Application: pressure flow of viscous media like polymer melts.

The use of dimensionless numbers in calculating non Newtonian flow problems is illustrated in Section 6.3.3 with an example.

■ 3.5 Heat Transfer by Convection

Heat transfer by convection, particularly forced convection, plays an important role in many polymer processing operations such as in cooling a blown film or a part in an injection mold, to mention only two examples. A number of expressions are found in the literature on heat transfer [3] for calculating the heat transfer coefficient, α, (see Section 3.1.6). The general relationship for forced convection has the form

$$Nu = f\left(Re, Pr\right) \tag{3.53}$$

The equation for the turbulent flow in a tube is given by [16]

$$Nu = 0.027 \, Re^{0.8} \cdot Pr \tag{3.54}$$

where

n = 0.4 for heating and

n = 0.3 for cooling

The following equation applies to laminar flow in a tube with a constant wall temperature [3]

$$Nu_{\text{lam}} = \sqrt[3]{3.66^3 + 1.61^3 \, Re \cdot Pr \cdot d_i \, / \, l} \tag{3.55}$$

where

d_i = inside tube diameter

l = tube length

The expression for the laminar flow heat transfer to a flat plate is [3]

$$Nu_{\text{lam}} = 0.664 \, \sqrt{Re} \, \sqrt[3]{Pr} \tag{3.56}$$

Equation (3.56) is valid for Pr = 0.6 to 2000 and $Re < 10^5$.

The equation for turbulent flow heat transfer to a flat plate is given as [3]

$$Nu_{\text{turb}} = \frac{0.037 \, Re^{0.8} \cdot Pr}{1 + 2.448 \, Re^{-0.1} \left(Pr^{2/3} - 1\right)} \tag{3.57}$$

Equation (3.57) applies for the following conditions:

Pr = 0.6 to 2000 and

$5 \cdot 10^5 < Re < 10^7$.

The properties of the fluids in the equations above are to be found at a mean fluid temperature.

Calculated Example

A flat film is moving in a coating equipment at a velocity of 130 m/min on rolls that are 200 mm apart. Calculate the heat transfer coefficient α if the surrounding medium is air at a temperature of 50 °C.

The properties of air at 50 °C are:

Kinematic viscosity v $= 17.86 \cdot 10^{-6}\, \text{m}^2/\text{s}$

Thermal conductivity λ $= 28.22 \cdot 10^{-3}\, \text{W}/(\text{m·K})$

Prandtl number $Pr = 0.69$

The Reynolds number Re_L, based on the length $L = 200$ mm is

$$Re_L = 130 \cdot \frac{1}{60} \cdot 0.2 \, / \, 17.86 \cdot 10^{-6} = 24{,}262$$

Substituting $Re_L = 24{,}262$ and $Pr = 0.69$ into Eq. (3.56) gives

$$Nu_{\text{lam}} = 0.664 \cdot 24{,}262^{0.5} \cdot 0.69^{1/3} = 91.53$$

As the fluid is in motion on both sides of the film, the Nusselt number is calculated according to [3]

$$Nu = \sqrt{Nu_{\text{lam}}^2 + Nu_{\text{turb}}^2}$$

or the turbulent flow Nu_{turb} follows from Eq. (3.57):

$$Nu_{\text{turb}} = \frac{0.037 \cdot 24{,}262^{0.8} \cdot 0.69}{1 + 2.448 \cdot 24{,}262^{-0.1} \left(0.69^{2/3} - 1\right)} = 102$$

The resulting Nusselt number Nu is

$$Nu = \sqrt{91.53^2 + 102^2} = 137$$

Heat transfer coefficient α results from

$$\alpha = \frac{Nu \cdot \lambda}{l} = \frac{137 \cdot 28.22 \cdot 10^{-3}}{0.2} = 19.33 \; \text{W}/(\text{m}^2 \cdot \text{K})$$

■ 3.6 Heat Transfer by Radiation

Heating by radiation is used in thermoforming processes to heat sheets or films so that the shaping process can take place. As at temperatures above 300 °C a substantial part of the thermal radiation consists of wavelengths in the infrared range, heat transfer by radiation is also termed infrared radiation [14]. According to the Stefan-Boltzmann law the rate of energy radiated by a black body per unit area $\dot{e}_s$ is proportional to the absolute temperature T to the fourth power (Fig. 3.13) [1]:

$$\dot{e}_s = \sigma\, T^4 \tag{3.58}$$

The Stefan-Boltzmann constant has the value

$$\sigma = 5.77 \cdot 10^{-12}\ \text{W/(cm}^2 \cdot \text{K}^4)$$

Equation (3.58) can also be written as

$$\dot{e}_s = c_s \left(\frac{T}{100}\right)^4 \tag{3.59}$$

where c_s = 5.77 W/(m²·K⁴)

The dependence of the black body radiation on the direction (Fig. 3.14) [1] is given by the cosine law of Lambert

$$\dot{e}_s = \dot{e}_n\, \cos\varphi \tag{3.60}$$

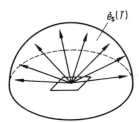

FIGURE 3.13 Blackbody radiation

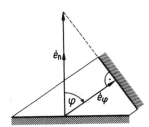

FIGURE 3.14 Lambert's law

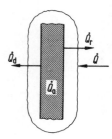

FIGURE 3.15 Properties of radiation

The radiation properties of technical surfaces are defined as (Fig. 3.15) [1]:

Reflectivity: $\rho \equiv \dfrac{\dot{Q}_r}{\dot{Q}}$ (3.61)

Absorptivity: $\alpha \equiv \dfrac{\dot{Q}_a}{\dot{Q}}$ (3.62)

Transmissivity: $\delta \equiv \dfrac{\dot{Q}_d}{\dot{Q}}$ (3.63)

The sum of these fractions must be unity, or

$\rho + \alpha + \delta = 1$

The transmissivity δ of opaque solids is zero so that

$\rho + \alpha = 1$

The reflectivity of gases ρ is zero and for those gases which emit and absorb radiation

$\alpha + \delta = 1$

Real bodies emit only a fraction of the radiant energy that is emitted by a black body at the same temperature. This ratio is defined as the emissivity ε of the body,

$\varepsilon = \dfrac{\dot{e}}{\dot{e}_s}$ (3.64)

At thermal equilibrium according to Kirchhoff's identity

$\varepsilon = \alpha$ (3.65)

Radiation Heat Transfer between Non-Black Surfaces

The net rate of radiant heat exchange between two infinite parallel plates is given by [15]

$$\dot{Q}_{12} = A \, \varepsilon_{12} \, c_s \left[\left(\frac{T_1}{100} \right)^4 - \left(\frac{T_2}{100} \right)^4 \right] \tag{3.66}$$

where

A = area

ε_{12} = emissivity factor defined by

$$\varepsilon_{12} = \frac{1}{\dfrac{1}{\varepsilon_1} + \dfrac{1}{\varepsilon_2} - 1} \tag{3.67}$$

Indices 1 and 2 refer to the two plates.

When T_2 is much smaller than T_1, the heat flow is approximately

$$\dot{Q}_{12} = \varepsilon_1 \, c_s \, A \left(\frac{T_1}{100} \right)^4 (0.1)$$

When heat transfer takes place by radiation and convection, the total heat transfer coefficient can be written as [15]

$$\alpha_{total} = \alpha_{convection} + \alpha_{radiation}$$

where

$$\alpha_{radiation} = \frac{\dot{Q}_{12}}{A \left(T_1 - T_2 \right)}$$

Calculated Example

A plastic sheet moving at a speed of 6 m/min is heated by two high temperature heating elements. Calculate the power required for heating for the following conditions, if the sheet is to be heated from 20 °C to 140 °C:

Net enthalpy of the plastic for a temperature difference of 120 °C: $\Delta h = 70$ kJ/kg

Width of the sheet	$w = 600$ mm
Thickness	$s = 250$ μm
Density of the resin	$\rho = 900$ kg/m³
Area of the heating element	$A = 0.0093$ m²
Emissivity of the heater	$\varepsilon = 0.9$

Solution:

Heating power N_H:

Mass flow rate of the plastic $\dot{m}$:

$$\dot{m} = U \cdot w \cdot s \cdot \rho = \frac{6 \cdot 0.6 \cdot 250 \cdot 900}{10^6} = 0.81 \, \text{kg/min}$$

$$N_H = \dot{m} \, \Delta h = \frac{0.81 \cdot 70 \cdot 1000}{60} = 945 \, \text{W}$$

As the area of the heating element is small compared to that of the sheet the equation shown below applies [14]

$$\dot{e} = \varepsilon \, c_s \cdot \left(\frac{T}{100}\right)^4 \, \text{W/m}^2$$

The total area of the heating element $A_g = 2 \cdot A$ so that we have

$$\dot{e} \, A_g = N_H$$

$$0.9 \cdot 5.77 \cdot 2 \cdot 0.0093 \cdot \left(\frac{T}{100}\right)^4 = 945$$

$$\frac{T}{100} = 9.95$$

$$T = 995 \, \text{K}$$

■ 3.7 Dielectric Heating

The dielectric heat loss that occurs in materials of low electrical conductivity when they are placed in an electric field alternating at high frequency is used in bonding operations, for instance, to heat seal plastic sheets or films.

The power dissipated in the polymer is given by [14]

$$N_{\mathrm{H}} = 2\,\pi\,f \cdot c \cdot E^2 \cdot \cot \phi \qquad (3.68)$$

where

N_{H} = power (W)

f = frequency of the alternating field (s^{-1})

C = capacitance of the polymer (farads)

E = applied voltage (Volt)

ϕ = phase angle

The rate of heat generation in a plastic film can be obtained from Eq. (3.68), and is given as [15]

$$N_{\mathrm{H}} = 55.7 \left(\frac{E^2 \cdot f \cdot \varepsilon_{\mathrm{r}}''}{4\,b^2} \right) \qquad (3.69)$$

where

N_{H} = rate of heat generation (W/m^3)

$\varepsilon_{\mathrm{r}}''$ = dielectric loss factor

b = half thickness of the film (μm)

Calculated Example

Given:

E = 500 V

f = 10 MHz

ε_r'' = 0.24

b = 50 µm

Calculate the rate of heat generation and the time required to heat the polymer from 20 °C to 150 °C.

Substituting the given values in Eq. (3.69) gives

$$N_H = \frac{55.7 \cdot 500^2 \cdot 10 \cdot 10^6 \cdot 0.24}{4 \cdot 50^2} = 3.34 \cdot 10^9 \ W/m^3$$

The maximum heating rate ΔT per second is calculated from

$$\Delta T = \frac{N_H}{c_p \cdot \rho} \qquad\qquad (3.70)$$

For

N_H = 3.39·109 W/m²

c_p = 2.2 kJ/(kg·K)

ρ = 800 kg/m³

$$\Delta T = \frac{3.39 \cdot 10^6}{800 \cdot 2.2} = 1926 \ K/s$$

Finally the heating time is

$$t = \frac{150 - 20}{1926} = 0.067 \ s$$

■ 3.8 Fick's Law of Diffusion

Analogous to Fourier's law of heat conduction (Eq. (3.31)) and the equation for shear stress in shear flow the diffusion rate in mass transfer is given by Fick's law. This can be written as [16]

$$\frac{\dot{m}_A}{A} = -D_{AB} \frac{\partial c_A}{\partial x} \qquad (3.71)$$

where

$\dot{m}_A$ = mass flux per unit time

A = Area

D_{AB} = diffusion coefficient of the constituent A in constituent B

c_A = mass concentration of component A per unit volume

x = distance

The governing expression for the transient rate of diffusion is [2]

$$\frac{\partial c_A}{\partial t} = D_{AB} \frac{\partial^2 c_A}{\partial x^2} \qquad (3.72)$$

where

t = time

x = distance

The desorption of volatile or gaseous components from a molten polymer in an extruder can be calculated from [17] using Eq. (3.72)

$$R_1 = A_c\, C_0 \sqrt{\frac{4 \cdot D}{\pi \cdot t}} \qquad (3.73)$$

where

R_1 = rate of desorption (g/s)

A_c = area of desorption (cm^2)

C_0 = initial concentration of the volatile component (g/cm^3) in the polymer

D = diffusion coefficient (cm^2/s)

t = time of exposure (s) of the polymer to the surrounding atmosphere

3.8.1 Permeability

Plastics are to some extent permeable to gases, vapors, and liquids. The diffusional characteristics of polymers can be described in terms of a quantity known as permeability.

The mass of the fluid permeating through the polymer at equilibrium conditions is given by [7]

$$m = \frac{P \cdot t \cdot A \cdot (p_1 - p_2)}{s} \tag{3.74}$$

where

m =   mass of the fluid permeating (g)

p =   permeability [g/(m·s·Pa)]

t =   time of diffusion (s)

A =   Area of the film or membrane (m^2)

p_1, p_2 =   partial pressures on the side 1 and 2 of the film (Pa)

s =   thickness of the film (m)

Besides its dependence on temperature, the permeability is influenced by the difference in partial pressures of the fluid and the thickness of the film. Other factors that influence permeability are the structure of the polymer film such as crystallinity and the type of fluid.

3.8.2 Absorption and Desorption

The process by which the fluid is absorbed or desorbed by a plastics material is time dependent and it is governed by its solubility and by the diffusion coefficient [7]. The period until its equilibrium value is reached can be very long. Its magnitude can be estimated by the half-life of the process given by [7]

$$t_{0.5} = \frac{0.04919 \cdot s^2}{D} \tag{3.75}$$

where

$t_{0.5}$ =   half life of the process

s =   thickness of the polymer assumed to be penetrated by one side

D =   diffusion coefficient

The value of $t_{0.5}$ for moisture in polymethyl methacrylate (PMMA) for

$$D = 0.3 \cdot 10^{-12} \text{ m}^2/\text{s} \quad \text{and} \quad s = 3 \text{ mm}$$

is 17.1 days, when the sheet is wet from one side only [7]. However, the equilibrium absorption takes much longer, as the absorption rate decreases with saturation.

■ 3.9 Case Study: Analyzing Air Gap Dynamics in Extrusion Coating by Means of Dimensional Analysis

A number of experimental investigations in the past have dealt with the influence of various parameters, such as melt temperature, air gap distance, and coating thickness on the bonding strength between a primed film and the coating. This section presents a general procedure for obtaining a quantitative relationship, in order to predict the influence of individual parameters on the bond strength on the basis of dimensional analysis. The applicability of this method was demonstrated by evaluating the experimental results published in the literature, which led to a practical formula for predicting adhesion as a function of the parameters mentioned.

It was found that for a given temperature the shear history and the residence time of the melt in the air gap are of utmost importance to the adhesion between the film and the coating. The application of the formulas given is explained by a number of numerical examples.

Laminating and coating processes involve a polymer film moving from a flat die at a high temperature and speed on to the rolls (Fig. 3.16). Experiments show that, once the film is successfully primed, the bond quality between the film and the coating depends on factors such as melt temperature, film thickness, and oxidation in the air gap, to mention a few [1].

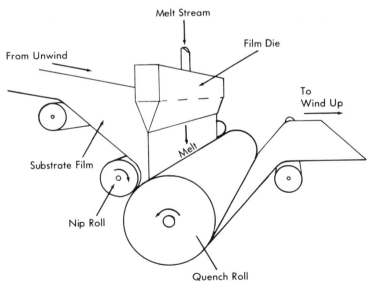

FIGURE 3.16 Extrusion coating process [25]

In this section, first the heat transfer between the film and the surrounding air is mathematically described, in order to predict the cooling of the film in the air gap. It was then shown that the bond strength depends on the shear history of the melt and the residence time of the film in the air gap for a given temperature by analyzing the experimental results on the basis of dimensional analysis.

3.9.1 Heat Transfer Between the Film and the Surrounding Air

The unsteady-state heat transfer between the film and the air by heat conduction can be calculated by means of the overall heat transfer coefficient [2] given by

$$k = \alpha_i = 6\,\lambda\,/\,s \tag{3.76}$$

k = overall heat transfer coefficient (W/m^2 K)

α_i = internal heat transfer coefficient (W/m^2 K)

λ = thermal conductivity (W/m·K)

s = film thickness (m)

Substituting relevant values for a LDPE film with a thickness of 20 μm, we get from Eq. (3.76) with λ = 0.24 W/m·K

$$\alpha_i = 6 \cdot 0.24 \,/\, 20 \cdot 10^{-6} = 72000 \text{ W/m}^2 \text{ K}$$

However, when the film moves in air, the external heat transfer coefficient α_a will be only of the order of 10 W/m^2 K [3] and determines the heat transfer between the film and the air. In other words, the heat from the film is controlled by the external resistance.

For this case, the temperature of the film in the air gap can be calculated from

$$\frac{T - T_\infty}{T_A - T_\infty} = \exp\left(-\frac{k \cdot A}{\rho \cdot c_p \cdot V} \cdot t\right) \tag{3.77}$$

where

T = final film temperature

T_∞ = air temperature

T_A = initial temperature of the film (melt temperature)

t = time in the air gap

Calculated Example

$T_A = 200\ °C,\quad T_\infty = 20\ °C,\quad k = \alpha_a = 10\ W/m^2\ K,\quad \rho = 700\ kg/m^3,\quad c_p = 2.3\ kJ/kg\ K,\quad$ and
$t = 200$ ms. Calculate the final film temperature.

Solution:

The ratio of the volume V to the heat transfer area A can be substituted by [2]

$$V / A = 2\ s \tag{3.78}$$

where s = film thickness.

The film thickness can be obtained from the coat weight cwt

$$s = 1000\ cwt/\rho \tag{3.79}$$

cwt   = coat weight (g/m^2)
ρ = density (kg/m^3)

For a coat weight of 10 g/m^2 the film thickness follows from Eq. (3.76) with ρ = 700 kg/m^3 for LDPE melt s = 14.28 µm.

Using a film thickness s = 30 µ and k = 10 $W/m^2\ K$,

the dimensionless expression $\dfrac{k \cdot A / V}{\rho \cdot c_p}$, called number of transfer of units, equals 0.02.

The right hand side of Eq. (3.76) equals 0.98 which leads to a final temperature of 196.4 °C for the conditions mentioned in the example. Thus, the cooling of the film in the air gap is not significant. However, at thicker films it can be considerable, and can affect the chemical kinetics of adhesion. In the evaluation of the experiments treated below, the final film temperature is assumed to be equal to the melt temperature. ∎

3.9.2 Chemical Kinetics

The effect of chemical kinetics on the adhesion was taken into account by means of the shift factor a_T [4, 5]

$$a_T = b_1 \cdot \exp\left[b_2 / (T + 273) \right] \tag{3.80}$$

Using the constants b_1 = 5.13 · 10^{-6} and b_2 = 5640 K, the shift factors at various melt temperatures were calculated.

Calculated Example

The shift factor at the melt temperature of $T = 321\ °C$ amounts to

$$a_T = 5.13 \cdot 10^{-6} \cdot \exp\left(5640 / 594\right) = 0.068$$

Time in the Air Gap (TIAG)

The Graetz number is defined as the ratio of the time to reach thermal equilibrium perpendicular to the flow direction to the residence time. The time in the air gap can be taken as the residence time and its effect on the bond strength predicted by means of the Graetz number Gz. This number can be calculated from [6]

$$Gz = \dot{m} \cdot c_p / \lambda \cdot GAPL \tag{3.81}$$

with

$\dot{m}$ = throughput (kg/h)

c_p = specific heat (kJ/kg K)

λ = thermal conductivity (W/m K)

$GAPL$ = Length of air gap (m)

Calculated Example

Calculate Gz for $m = 197$ kg/h, $c_p = 2.2$ kJ/kg K, $\lambda = 2.2$ W/m K, and $GAPL = 102$ mm. Using coherent dimensions Gz follows from Eq. (3.81)

$$Gz = 10 \cdot 197 \cdot 2.2 / 36 \cdot 0.24 \cdot 0.102 = 4918$$

Shear History of the Film

It was found [1] that the shear history of the film plays a significant role in the gap dynamics. The shear can be calculated by multiplying the shear rate of the film at the die exit and time in the air gap (TIAG).

$$\text{SHEAR} = \text{Shear rate at the die exit} \times \text{TIAG}$$

3.9.3 Evaluation of the Experiments

The experimental results of Ristey and Schroff [23] were evaluated using the concepts described above. The Graetz numbers calculated from the experiments lie in the range of 1600 to 5000. A shear rate range of 69 s^{-1} to 208 s^{-1} was taken as a basis.

By applying stepwise linear regression, the following equation has been developed from the experimental results

$$\ln(C = 0) = -6.7684 - 3.1784 \ln(\text{shift factor}) - 0.8487 \ln(Gz) + 1.2784 \ln(\text{shear}) \qquad (3.82)$$

with $C = 0$: Carbonyl Absorbance.

The 45 degree slope in Fig. 3.17 shows the calculated and experimental results. Owing to the complicated processes involved in the air gap dynamics, there is some deviation between calculations and measurements.

Still, it is seen that Eq. (3.82) predicts the combined effect of the significant parameters not only in the trend but in quantitative terms as well close to reality.

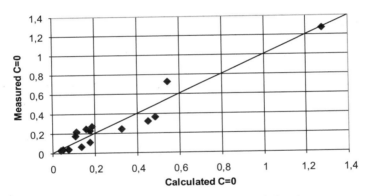

FIGURE 3.17 Comparison between measured [6] and calculated carbonyl absorbance

■ References

[1]   Bender, E.: Lecture notes, Wärme and Stoffübergang, Univ. Kaiserslautern (1982)
[2]   McCabe, W. L.; Smith, J. C.; Harriott, P.: Unit Operations of Chemical Engineering. McGraw Hill, New York (1985)
[3]   Martin, H.: in VDI Wärmeatlas, VDI Verlag, Düsseldorf (1984)
[4]   Welty, J. R.; Wicks, C. E.; Wilson, R. E.: Fundamentals of Momentum, Heat and Mass Transfer, John Wiley, New York (1983)
[5]   Kreith, F.; Black, W. Z.: Basic Heat Transfer, Harper & Row, New York (1980)
[6]   Thorne, J. L.: Plastics Process Engineering, Marcel Dekker Inc., New York (1979)
[7]   Ogorkiewicz, R. M.: Thermoplastics – Properties and Design, John Wiley, New York (1974)
[8]   Wübken, G.: Berechnen von Spritzgießwerkzeugen, VDI Verlag, Düsseldorf (1974)
[9]   Gersten, K.: Einführung in die Strömungsmechanik, Vieweg, Braunschweig (1981)
[10] Winter, H. H.: Extruder als Plastifiziereinheit, VDI Verlag, Düsseldorf (1977)
[11] Rauwendaal, C.: Polymer Extrusion 4th ed., Hanser, Munich (2001)
[12] Kremer, H.: Grundlagen der Feuerungstechnik, Engler-Bunte-Institute, Univ. Karlsruhe (1964)
[13] Cogswell, F. N.: Polymer Melt Rheology, George Godwin Ltd., London (1981)
[14] Bernhardt, E. C.: Processing of Thermoplastic Materials, Reinhold, New York (1959)
[15] McKelvey, J. M.: Polymer Processing, John Wiley, New York (1962)
[16] Holman, J. P.: Heat Transfer, McGraw Hill, New York (1980)
[17] Secor, R. M.: *Polymer Engineering and Science*, **26** (1986) p. 647
[18] Foster, B. W.: Proc. TAPPI, PLACE, Boston (2002)
[19] Martin, H.: Heat Exchangers, Hemisphere Publishing Corporation, PA, USA(1992)
[20] McCabe, W. L.; Smith, J. C.; Harriott, P.: Unit Operations of Chemical Engineering, McGraw Hill (1985)
[21] Muenstedt, H.: *Kunststoffe*, **68**, (1978), p. 92
[22] Rao, N.S : Design Formulas for Plastics Engineers, Hanser, Munich (1991)
[23] Ristey, W. J.; Schroff, R. N.: TAPPI Paper Synthetics Proceedings, TAPPI Press, Atlanta (1978), p. 267
[24] Rao, N. S.; Schumacher, G.; Ma, C. T.: *J. Injection Molding Technology*, **4** (2000), p. 92
[25] Osborne, K. R.; Jenkins, W. A.: Plastic Films, Technomic Publishing, Lancaster (1992)

4 Analytical Procedures for Troubleshooting Extrusion Screws

Extrusion is one of the most widely used polymer converting operations for manufacturing blown film, pipes, sheets, and laminations, to list the most significant industrial applications. Fig. 4.1 shows a modern large scale machine for making blown film. The extruder, which constitutes the central unit of these machines, is shown in Fig. 4.2. The polymer is fed into the hopper in the form of granulate or powder. It is kept at the desired temperature and humidity by controlled air circulation. The solids are conveyed by the rotating screw and slowly melted, in part, by barrel heating but mainly by the frictional heat generated by the shear between the polymer and the barrel (Fig. 4.3). The melt at the desired temperature and pressure flows through the die, in which the shaping of the melt into the desired shape takes place.

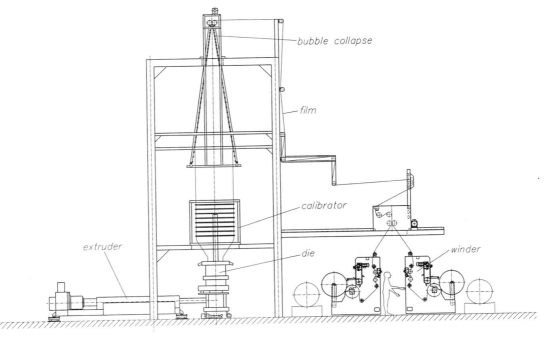

FIGURE 4.1 Large scale blown film line [2]

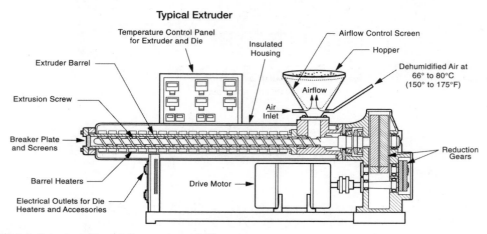

FIGURE 4.2 Extruder with auxialiary equipment [3]

■ 4.1 Three-Zone Screw

Basically extrusion consists of transporting the solid polymer in an extruder by means of a rotating screw, melting the solid, homogenizing the melt, and forcing the melt through a die (Fig. 4.3).

The extruder screw of a conventional plasticating extruder has three geometrically different zones (Fig. 4.4), whose functions can be described as follows:

Feed zone: Transport and preheating of the solid material

Transition zone: Compression and plastication of the polymer

Metering zone: Melt conveying, melt mixing and pumping of the melt to the die

However, the functions of a zone are not limited to that particular zone alone. The processes mentioned can continue to occur in the adjoining zone as well.

Although the following equations apply to the 3-zone screws, they can be used segmentwise for designing screws of other geometries as well.

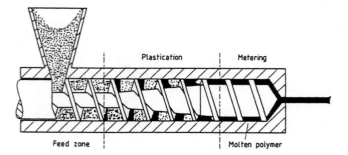

FIGURE 4.3 Plasticating extrusion [4]

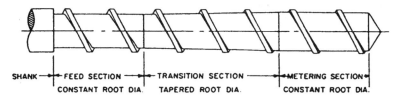

SHANK — FEED SECTION — TRANSITION SECTION — METERING SECTION
CONSTANT ROOT DIA.   TAPERED ROOT DIA.   CONSTANT ROOT DIA.

FIGURE 4.4 Three-zone screw [10]

4.1.1 Extruder Output

Depending on the type of extruder, the output is determined either by the geometry of the solids feeding zone alone, as in the case of a grooved extruder [7], or by the solids and melt zones to be found in a smooth barrel extruder. A too high or too low output results when the dimensions of the screw and die are not matched with each other.

4.1.2 Feed Zone

A good estimate of the solids flow rate can be obtained from Eq. (4.1) as a function of the conveying efficiency and the feed depth. The desired output can be found by simulating the effect of these factors on the flow rate by means of Eq. (4.1).

Calculated Example

The solids transport is largely influenced by the frictional forces between the solid polymer and barrel and screw surfaces. A detailed analysis of the solids conveying mechanism was performed by Darnell and Mol [8]. The following example presents an empirical equation that provides good results in practice [1].

The geometry of the feed zone of a screw (Fig. 4.5) is given by the following data:

Barrel diameter	D_b	= 30 mm
Screw lead	s	= 30 mm
Number of flights	ν	= 1
Flight width	w_{FLT}	= 3 mm
Channel width	W	= 28.6 mm
Depth of the feed zone	H	= 5 mm
Conveying efficiency	η_F	= 0.436
Screw speed	N	= 250 rpm
Bulk density of the polymer	ρ_o	= 800 kg/m^3

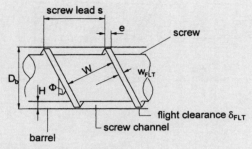

FIGURE 4.5 Screw zone of a single screw extruder [5]

The solids conveying rate in the feed zone of the extruder can be calculated according to [4]

$$G = 60 \cdot \rho_o \cdot N \cdot \eta_F \cdot \pi^2 \cdot H \cdot D_b \left(D_b - H\right) \frac{W}{W + w_{FLT}} \cdot \sin\phi \cdot \cos\phi \qquad (4.1)$$

with the helix angle ϕ

$$\phi = \tan^{-1}\left[s / \left(\pi \cdot D_b\right)\right] \qquad (4.2)$$

The conveying efficiency η_F in Eq. (4.1) as defined here is the ratio between the actual extrusion rate and the theoretical maximum extrusion rate attainable under the assumption of no friction between the solid polymer and the screw. It depends on the type of polymer, bulk density, barrel temperature, and the friction between the polymer, barrel and the screw. Experimental values of η_F for some polymers are given in Table 4.1.

TABLE 4.1 Conveying efficiency η_F for some polymers

Polymer	Smooth barrel	Grooved barrel
LDPE	0.44	0.8
HDPE	0.35	0.75
PP	0.25	0.6
PVC-P	0.45	0.8
PA	0.2	0.5
PET	0.17	0.52
PC	0.18	0.51
PS	0.22	0.65

Using the values above with the dimensions in meters in Eq. (4.1) and Eq. (4.2) we get

$$G = 60 \cdot 800 \cdot 250 \cdot 0.44 \cdot \pi^2 \cdot 0.005 \cdot 0.03 \cdot 0.025 \cdot \frac{0.0256}{0.0286} \cdot 0.3034 \cdot 0.953$$

Hence $G \approx 50$ kg/h

4.1.3 Metering Zone (Melt Zone)

The conveying capacity of the melt zone must correspond to the amount of melt created by plastication, in order to ensure stable melt flow without surging. This quantity can be estimated by means of the Eq. (4.8).

Starting with the parallel plate model and correcting it by means of appropriate correction factors [5], the melt conveying capacity of the metering zone can be calculated [5]. Although the following equations are valid for an isothermal quasi-Newtonian fluid, they were found to be useful for many practical applications [1].

These equations can be summarized as follows:

Volume flow rate of pressure flow $\dot{Q}_p$:

$$\dot{Q}_p = \frac{-\pi \, D_b \cdot H^3 \left(1 - \dfrac{v \cdot e}{s}\right) \cdot \sin^2 \varphi \cdot \Delta p \cdot 10^{-4}}{12 \, \eta_a \cdot L} \tag{4.3}$$

Mass flow rate $\dot{m}_p$:

$$\dot{m}_p = 3600 \cdot 1000 \cdot \dot{Q}_p \cdot \rho_m \tag{4.4}$$

Drag flow $\dot{Q}_d$ (Fig. 4.5):

$$\dot{Q}_d = \frac{\pi^2 \cdot D_b^2 \cdot N \cdot H \cdot \left(1 - \dfrac{v \cdot e}{s}\right) \cdot \sin \varphi \cdot \cos \varphi \cdot 10^{-9}}{120} \tag{4.5}$$

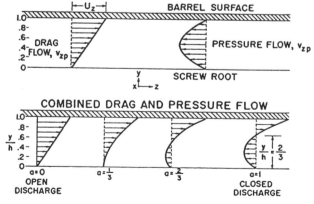

FIGURE 4.6 Drag and pressure flow in screw channel [5]

Mass flow rate $\dot{m}_d$:

$$\dot{m}_d = 3600 \cdot 1000 \cdot \dot{Q}_d \cdot \rho_m \tag{4.6}$$

Leakage flow $\dot{Q}_L$:

To avoid metal to metal friction, extrusion screws have a small clearance between the top of the flight and the barrel surface. This clearance reduces the pumping rate of the melt because it enables the polymer to leak across the flights. The net flow rate $\dot{Q}$ is therefore

$$\dot{Q} = \dot{Q}_d + \dot{Q}_p - \dot{Q}_L \tag{4.7}$$

The melt conveying rate of the metering zone can be calculated from [1]

$$\dot{m} = 3 \cdot 10^{-5} \cdot \pi^2 \cdot D_b^2 \cdot N \cdot H \cdot \left(1 - \frac{v \cdot e}{s}\right) \cdot \rho_m \cdot \sin\phi \cdot \cos\phi \cdot \left(1 - a_d - J\right) \tag{4.8}$$

where $a_d = -\dot{m}_p / \dot{m}_d$ and $J = \delta_{FLT} / H$

The average shear rate in the metering channel is calculated as

$$\dot{\gamma}_a = \pi \cdot D_b \cdot N / 60 \cdot H \tag{4.9}$$

Symbols and units used in the equations above are

D_b	Barrel diameter [mm]
H	Channel depth [mm]
e	Flight width [mm]
s	Screw lead [mm]
δ_{FLT}	Flight clearance [mm]
L	Length of metering zone [mm]
$\dot{Q}_p, \dot{Q}_d$	Volume flow rate of pressure and drag flow, respectively [m³/s]
$\dot{m}_p, \dot{m}_d$	Mass flow rate of pressure and drag flow, respectively [kg/h]
$\dot{m}$	Extruder output [kg/h]
Δp	Pressure difference across the metering zone [bar]
v	Number of flights
η_a	Melt viscosity [Pa·s]
$\dot{\gamma}_a$	Average shear rate [s⁻¹]
ρ_m	Density of the melt [g/cm³]
a_d	Ratio of pressure flow to drag flow
N	Screw speed [rpm]

Calculated Example

For the following conditions the extruder output is to be determined:

Melt viscosity η_a = 1406.34 Pa·s; N = 80 rpm; Δp = 300 bar; ρ_m = 0.7 g/cm³; D_b = 60 mm; H = 3 mm; e = 6 mm; s = 60 mm; δ_{FLT} = 0.1 mm; L = 600 mm; v = 1

The symbols above refer to Fig. 4.5.

Substituting these values into the equations above one obtains

$$\dot{m}_p = 3.148 \text{ kg/h} \qquad\qquad \text{Eq. (4.4)}$$

$$\dot{m}_d = 46.59 \text{ kg/h} \qquad\qquad \text{Eq. (4.6)}$$

$$\dot{m} = 41.88 \text{ kg/h} \qquad\qquad \text{Eq. (4.8)}$$

Leakage flow

$$\dot{m}_l = \dot{m}_d + \dot{m}_p - \dot{m} = 1.562 \text{ kg/h}$$

With the help of Eq. (4.8) the effect of different parameters on the extruder output is presented in Figs. 4.7 to 4.14 by changing one variable at a time and keeping all other variables constant. The dimensions of the screw used in these calculations are taken from the example above.

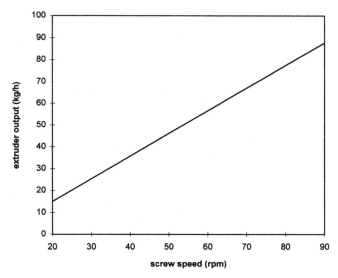

FIGURE 4.7 Effect of screw speed on extruder output

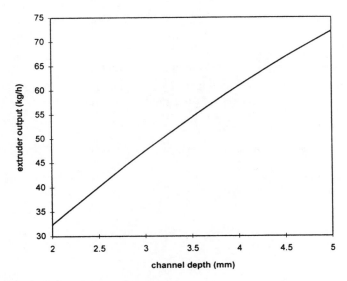

FIGURE 4.8 Effect of channel depth on extruder output

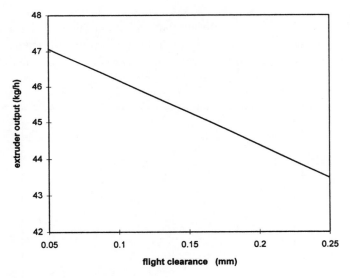

FIGURE 4.9 Effect of flight clearance on extruder output

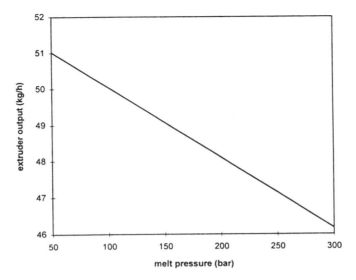

FIGURE 4.10 Effect of melt pressure on extruder output

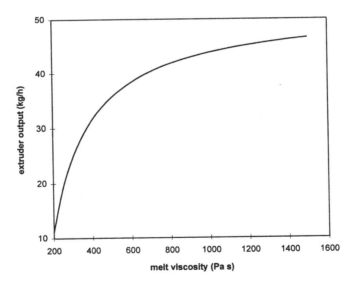

FIGURE 4.11 Effect of melt viscosity on extruder output

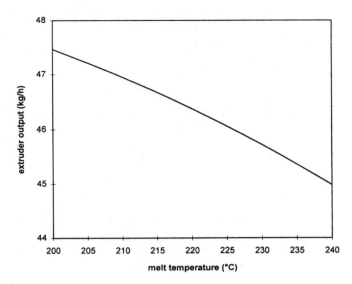

FIGURE 4.12 Effect of melt temperature on extruder output

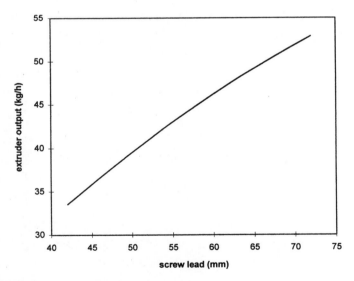

FIGURE 4.13 Effect of screw lead on extruder output

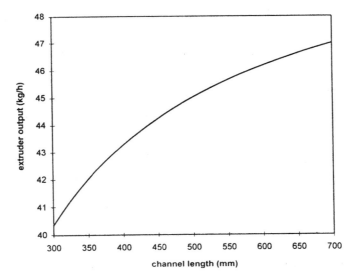

FIGURE 4.14 Effect of channel length on extruder output

In a particular troubleshooting case, surging was avoided by increasing the length of the metering zone. The means of stabilizing the melt flow vary from case to case, and can be determined from the simulations, as shown above. These simulations can be carried out easily with the help of the computer program VISOUT, listed in the Appendix.

Correction Factors

To correct the infinite parallel plate model for the flight edge effects the following factors can be used along with the equations above:

With sufficient accuracy the shape factor F_d for the drag flow can be obtained from [10]

$$F_d = 1 - 0.571 \frac{H}{W} \tag{4.10}$$

and the factor F_p for the pressure flow

$$F_p = 1 - 0.625 \frac{H}{W} \tag{4.11}$$

The expressions for the corrected drag flow and pressure flow would then be

$$\dot{Q}_{ab} = F_d \cdot \dot{Q}_d \tag{4.12}$$

and

$$\dot{Q}_{pk} = F_p \cdot \dot{Q}_p \tag{4.13}$$

The correction factor for the screw power, which is treated in the next section, can be determined from [5]

$$F_x = e^x - x^3 + 2.2\,x^2 - 1.05\,x \tag{4.14}$$

with $x = H/W$.

Equation (4.14) is valid in the range $0 < H/W < 2$. For the range of commonly occurring H/W-ratios in extruder screws, the flight edge effect accounts for only less than 5% and can therefore be neglected [10]. The influence of screw curvature is also small so that F_x can be taken as 1. Although the above mentioned factors are valid only for Newtonian fluids, their use for polymer melt flow is accurate enough for practical purposes.

4.1.4 Practical Design of 3-Zone Screws

Based on the laws of similarity Pearson [16] developed a set of relationships to scale-up a single screw extruder. These relationships are useful for the practicing engineer to estimate the size of a larger extruder from experimental data gathered on a smaller machine. The scale-up assumes equal length to diameter ratios between the two extruders. The important relations can be summarized as follows:

$$\frac{H_2}{H_1} = \left(\frac{D_2}{D_1}\right)^{(1-s)/(2-3s)} \tag{4.15}$$

$$\frac{N_2}{N_1} = \left(\frac{D_2}{D_1}\right)^{-(2-2s)/(2-3s)} \tag{4.16}$$

$$\frac{\dot{m}_2}{\dot{m}_1} = \left(\frac{D_2}{D_1}\right)^{(3-5s)/(2-3s)} \tag{4.17}$$

$$\frac{H_{F_2}}{H_{F_1}} = \left(\frac{D_2}{D_1}\right)^{(1-s)/(2-3s)} \tag{4.18}$$

where

H_F = feed depth

H = metering depth

D = screw diameter

N = screw speed

Indices 1, 2 = screw of known geometry and screw for which the geometry is to be determined, respectively.

The exponent s is given by

$$s = 0.5 \left(1 - n_{\mathrm{R}}\right) \tag{4.19}$$

where n_{R} is the reciprocal of the power law exponent n. The shear rate required to determine n is obtained from

$$\dot{\gamma}_a = \frac{\pi \cdot D_1 \cdot N_1}{60 \cdot H_1} \tag{4.20}$$

Calculated Example

The following conditions are given:

The resin is LDPE with the same constants of viscosity as in the Calculated Example in Section 4.1.1.2. The stock temperature is 200 °C. The data pertaining to screw 1 are:

D_1 = 90 mm; H_F = 12 mm; H_1 = 4 mm

Feed length   = $9\,D_1$

Transition length = $2\,D_1$

Metering length   = $9\,D_1$

Output $\dot{m}_1$ = 130 kg/h

Screw speed N_1 = 80 rpm

The diameter of screw 2 is D_2 = 120 mm. The geometry of screw 2 is to be determined.

Solution:

The geometry is computed from the equations given above [1]. It follows that

D_2 = 120 mm

H_{F_2} = 14.41 mm

H_2 = 4.8 mm

$\dot{m}_2$ = 192.5 kg/h

N_1 = 55.5 rpm

Other methods of scaling up have been treated by Schenkel [18], Fischer [19], and Potente [20]. Examples for calculating the dimensions of extrusion screws are given below.

Calculated Example

Determine the specific output for a LLDPE resin with a power law exponent $n = 2$.

Solution:
Substituting $n = 2$ into Eq. (4.19) we get $s = 0.25$. Using Eqs. (4.16) and (4.17), we finally obtain

$$\dot{m}_s = \frac{\dot{m}}{N} = 0.0000155 \, D^{2.6} \qquad\qquad (4.21)$$

where
$\dot{m}$ = output kg/h
N = screw speed rpm
D = screw diameter mm
$\dot{m}_s$ = specific output

The specific output $\dot{m}_s$ is plotted as a function of the screw diameter D for LLDPE and LDPE in Fig. 4.15 by means of Eq. (4.21) and the formula derived for LDPE with the exponent $n = 2.5$ from Eqs. (4.16), (4.17) and (4.19).

Other target values of practical interest are plotted in Fig. 4.16 to Fig. 4.19 as a function of screw diameter. Machine manufacturers confirmed the application of these easy-to-use relationships, although only one single value of the power law exponent is used to characterize the resin flow.

The scale-up of extrusion screws can be performed quickly by using the program VISSCALE given in the Appendix.

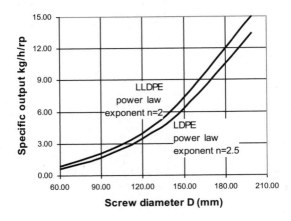

FIGURE 4.15 Specific output $\dot{m}_s$ for LDPE and LLDPE as a function of screw diameter for an extruder of $L/D = 20$

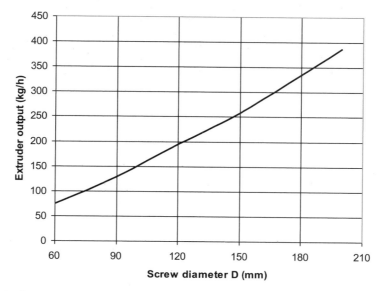

FIGURE 4.16 Extruder output vs. screw diameter for LDPE and extruder of $L/D = 20$, $\dot{m} = 0.281 \cdot D^{1.364}$ kg/h

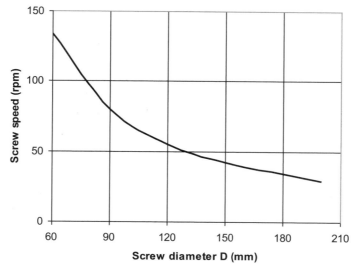

FIGURE 4.17 Screw speed vs. screw diameter for LDPE and extruder of $L/D = 20$, $N = 24600\, D^{-1.273}$ rpm

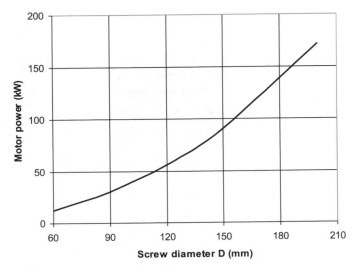

FIGURE 4.18 Motor power vs. screw diameter for LDPE and extruder of $L/D = 20$, PWR $= 0.0015 \cdot D^{2.2}$ kW

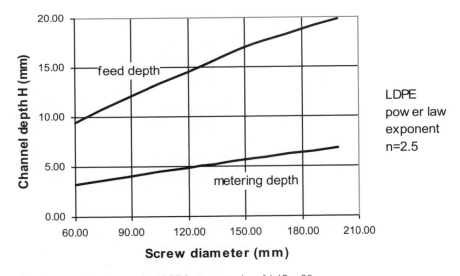

LDPE
pow er law
exponent
n=2.5

FIGURE 4.19 Channel depths of the screw for LDPE for an extruder of $L/D = 20$

■ 4.2 Melting of Solids

Physical models describing the melting of solids in extruder channels were developed by many workers, notable among them the work of Tadmor [4]. Rauwendaal summarizes the theories underlying these models in his book [10]. Detailed computer programs for calculating melting profiles based on these models have been given by Rao in the books [1, 11] and in the Appendix.

The purpose of the following section is to illustrate the calculation of the main parameters occurring in these models through numerical examples. The important steps for obtaining a melting profile are treated in the Section 4.2.2.

4.2.1 Thickness of Melt Film

According to the Tadmor model [4], the maximum thickness of the melt film (Fig. 4.20) is given by

$$\delta_{max} = \left\{ \frac{\left[2\,\lambda_m \left(T_b - T_m \right) + \eta_f\, V_j^2 \cdot 10^{-4} \right] W}{10^3 \cdot V_{bx} \cdot \rho_m \left[c_{ps} \left(T_m - T_s \right) + i_m \right]} \right\}^{0.5} \qquad (4.22)$$

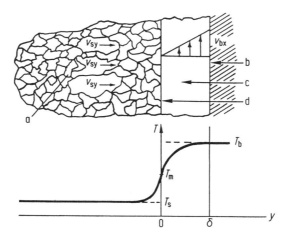

FIGURE 4.20 Temperature profile in the melt film after Tadmor [4];
a: solid bed, b: barrel surface, c: melt film, d: solid melt interface [4]

Example with Symbols and Units

Thermal conductivity of the melt $\lambda_m = 0.174\ \text{W/(mK)}$

Barrel temperature $T_b = 150\ ^\circ\text{C}$

Melting point of the polymer $T_m = 110\ ^\circ\text{C}$

Viscosity in the melt film $\eta_f\ (\text{Pa·s})$

Shear rate in the film $\dot{\gamma}_f\ (\text{s}^{-1})$

Velocity of the barrel surface $V_b\ (\text{cm/s})$

Velocity components $V_{bx},\ V_{bz}\ (\text{cm/s})$, see Fig. 4.21

Velocity of the solid bed $V_{sz}\ (\text{cm/s})$

Output of the extruder $G = 16.67\ \text{g/s}$

Average film thickness $\bar{\delta}_f\ \text{mm}$

Temperature of the melt in the film $\bar{T}_f\ (^\circ\text{C})$

Average film temperature $T_a\ (^\circ\text{C})$

Depth of the feed zone $H_1 = 9\ \text{mm}$

Width of the screw channel $W = 51.46\ \text{mm}$

Melt density $\rho_m = 0.7\ \text{g/cm}^3$

Density of the solid polymer $\rho_s = 0.92\ \text{g/cm}^3$

Specific heat of the solid polymer $c_{ps} = 2.2\ \text{kJ/(kg·K)}$

Temperature of the solid polymer $T_s = 20\ ^\circ\text{C}$

Heat of fusion of the polymer $i_m = 125.5\ \text{kJ/kg}$

Maximum film thickness $\delta_{max}\ (\text{cm})$

Indices: m: melt; s: solid

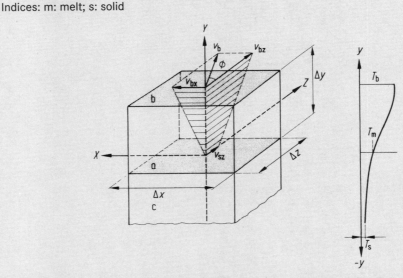

FIGURE 4.21 Velocity and temperature profiles in the melt and solid bed after Tadmor [4]; a: solid melt interface, b: cylinder, c: solid bed

Calculated Example

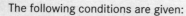

The following conditions are given:

The resin is LDPE with the same constants of viscosity as in the example of Section 4.1.1.2.

The barrel diameter D_b is 60 mm and the screw speed is 80 rpm.

$$V_b = \frac{\pi \cdot D_b \cdot N}{10 \cdot 60} = \frac{\pi \cdot 60 \cdot 80}{10 \cdot 60} = 25.13 \text{ cm/s}$$

$$V_{bx} = V_b \sin\phi = 25.1 \cdot \sin 17.66° = 7.62 \text{ cm/s}$$

$$V_{bz} = V_b \cos\phi = 25.1 \cdot \cos 17.66° = 23.95 \text{ cm/s}$$

$$V_{sz} = \frac{10 \cdot G}{W \cdot H_1 \cdot \rho_s} = \frac{100 \cdot 16.67}{51.46 \cdot 9 \cdot 0.92} = 3.91 \text{ cm/s}$$

Relative velocity V_j (Fig. 4.21):

$$V_j = \left(V_b^2 + V_{sz}^2 - 2 V_b \cdot V_{sz} \cdot \cos\phi\right)^{0.5}$$

$$= \left(25.13^2 + 3.91^2 - 2 \cdot 25.31 \cdot 3.91 \cdot \cos 17.66°\right)^{0.5}$$

$$= 21.44 \text{ cm/s}$$

Temperature $\overline{T}_a$:

$$\overline{T}_a = \frac{T_b + T_m}{2} = \frac{150 + 110}{2} = 130 \text{ °C}$$

$$\dot{\gamma}_a = \frac{V_j}{\delta_f}$$

Starting from an assumed film thickness of 0.1 mm and using the temperature resulting when heat generation is neglected, the viscosity in the film is first estimated. By changing the film thickness and repeating this calculation, the final viscosity is obtained [1].

This iteration leads to

$$\dot{\gamma}_f = \frac{V_j}{\delta} = \frac{2 \cdot 21.44 \cdot 10}{0.299} = 1434 \text{ s}^{-1}$$

$$\eta_f = 351.84 \text{ Pa} \cdot \text{s}$$

δ_{max} from Eq. (4.22):

$$\delta_{max} = \left\{ \left[2 \cdot 0.174 \left(150 - 110 \right) + 351.84 \cdot 21.44^2 \cdot 10^{-4} \right] \cdot \frac{51.46}{10^3 \cdot 7.62 \cdot 0.7 \left[2.2 \left(110 - 20 \right) + 125.5 \right]} \right\}$$

$$= 0.0299 \text{ cm}, \quad \text{or} \quad 0.299 \text{ mm}$$

∎

Temperature in the Melt Film

Taking the viscous heat generation into account the temperature in melt film can be obtained from [4]

$$\overline{T_f} = \overline{T_a} + \frac{10^{-4} \cdot \eta_f \cdot V_j^2}{12 \cdot \lambda_m} \tag{4.23}$$

$$\overline{T_f} - \overline{T_a} = \frac{10^{-4} \cdot 351.84 \cdot 21.44^2}{12 \cdot 0.174} = 7.75 \,^\circ\text{C}$$

$$\overline{T_f} = \left(\frac{150 + 110}{2} \right) + 7.75 = 137.5 \,^\circ\text{C}$$

As seen from the equations above, the desired quantities have to be calculated on an iterative basis. This is done by the computer program TEMPMELT mentioned in the Appendix.

Melting Rate

The melting rate is described by Tadmor [4] through the parameter ϕ_p, which is expressed as

$$\phi_p = \left\{ \frac{V_{bx} \cdot \rho_m \cdot \left[\lambda_m \left(T_b - T_m \right) + 0.5 \, \eta_f \cdot V_j^2 \cdot 10^{-4} \right]}{100 \cdot 2 \left[c_{ps} \left(T_m - T_s \right) + i_m \right]} \right\}^{0.5} \tag{4.24}$$

The numerator represents the heat supplied to the polymer by conduction through the barrel and dissipation, whereas the denominator shows the enthalpy required to melt the solid polymer. The melting rate increases with increasing ϕ_p.

By inserting the known parameters into Eq. (4.24) we obtain

$$\phi_p = \left\{ \frac{7.62 \cdot 0.7 \left[0.174 \left(150 - 110 \right) + 0.5 \cdot 351.84 \cdot 21.44^2 \cdot 10^{-4} \right]}{100 \cdot 2 \left[2.2 \left(110 - 20 \right) + 125.5 \right]} \right\}^{0.5} = 0.035 \, \frac{\text{g}}{\text{cm}^{1.5} \, \text{s}}$$

Dimensionless Melting Parameter

The dimensionless melting parameter ψ is defined as [4]

$$\psi = \frac{\phi_p \cdot H_1 \cdot W^{0.5}}{10^{1.5} \cdot G} \qquad (4.25)$$

with

ϕ_p = 0.035 g/(cm$^{1.5}$·s)

H_1 = 9 mm

W = 51.46 mm and

G = 16.67 g/s

we get

$$\psi = 0.004$$

The dimensionless parameter is the ratio between the amount of melted polymer per unit down channel distance to the extruder output per unit channel feed depth.

4.2.2 Melting Profile

The melting profile gives the amount of unmelted polymer as a function of screw length (Fig. 4.22) and is the basis for calculating the stock temperature and pressure. It thus shows whether the polymer at the end of the screw is fully melted. The plasticating and mixing capacity of a screw can be improved by mixing devices. Knowledge of the melting profile enables one to judge the suitable positioning of mixing and shearing devices in the screw [17].

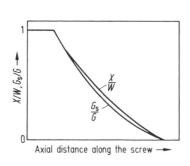
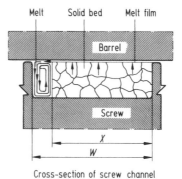

FIGURE 4.22 Solid bed or melting profiles X/W and G_s/G [4, 17]
with G: total mass flow rate, G_s: mass flow rate of solids

The following equation applies to a screw zone of constant depth [4]

$$\frac{X_{out}}{W} = \frac{X_{in}}{W}\left(1 - \frac{\psi \cdot \Delta z}{2 \cdot H_1}\right)^2 \tag{4.26}$$

and for a tapered channel [4]

$$\frac{X_{out}}{W} = \frac{X_{in}}{W}\left[\frac{\psi}{A} - \left(\frac{\psi}{A} - 1\right)\sqrt{\frac{H_{in}}{H_{out}}}\right]^2 \tag{4.27}$$

where

$$A = \frac{H_1 - H_2}{Z} \tag{4.28}$$

The parameter ψ is obtained from Eq. (4.25).

Symbols and units:

X_{out}, X_{in} mm   Width of the solid bed at the outlet and inlet of a screw increment, respectively

W mm   Channel width

ψ Melting parameter

Δz mm   Downchannel distance of the increment

H_{in}, H_{out} mm   Channel depth at the inlet and outlet of an increment, respectively

H_1, H_2 mm   Channel depth of a parallel zone (feed zone) and depth at the end of a transition zone (Fig. 4.23)

A Relative decrease of channel depth, Eq. (4.28)

Z mm   Downchannel length of a screw zone

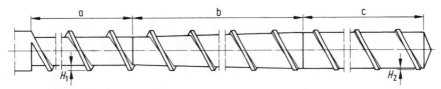

FIGURE 4.23 Three-zone screw [10]

Calculated Example

a) Constant channel depth

For $H_1 = 9$ mm; $X_{in}/W = 1$; $\Delta z = 99$ mm, and $\psi = 0.004$ from Section 4.2.1, X_{out}/W can be calculated from Eq. (4.26):

$$\frac{X_{out}}{W} = 1 \left(1 - \frac{0.004 \cdot 99}{2 \cdot 9} \right)^2 = 0.96$$

This means that at a distance of $\Delta z = 99$ mm, 4% of the solids were melted.

b) Varying channel depth

For the values

H_1 = 9 mm

H_2 = 3 mm

Z = 1584 mm

X_{in}/W = 0.96

H_{in} = 9 mm

H_{out} = 8.625 mm

ψ can be obtained from Eq. (4.25)

$$\psi = \frac{\phi_p \cdot H_1 \cdot W^{0.5}}{10^{1.5} \cdot \left(\dfrac{X_{in}}{W} \right) \cdot G} = \frac{0.035 \cdot 9 \cdot 51.46^{0.5}}{10^{1.5} \cdot 0.96 \cdot 16.67} = 0.00447$$

The relative decrease of the channel depth A is calculated from Eq. (4.28):

$$A = \frac{H_1 - H_2}{Z} = \frac{9 - 3}{1584} = 0.00379$$

and X_{out}/W from Eq. (4.27)

$$\frac{X_{out}}{W} = 0.96 \left[\frac{0.0047}{0.00379} - \left(\frac{0.0047}{0.00379} - 1 \right) \sqrt{\frac{9}{8.625}} \right]^2$$

Assuming a constant velocity of the solid bed, the mass flow ratio G_s/G results from

$$\frac{G_s}{G} = \frac{\overline{X}\,\overline{H}}{W\,H_1} \tag{4.29}$$

where

G_S = mass flow rate of the solid polymer (g/s)

G = throughout of the extruder (g/s)

$\overline{X}$ = average of X_{in} and X_{out} (mm)

$\overline{H}$ = average of H_{out}, and H_{in} (mm)

For a zone of constant depth it follows that

$$\frac{G_s}{G} = \frac{\overline{X}}{W} \tag{4.30}$$

a) Constant depth

$$\frac{\overline{X}}{W} = 0.5 \left(\frac{X_{in}}{W} + \frac{X_{out}}{W} \right) = 0.5 \cdot 1.96 = 0.98$$

$$\frac{G_s}{G} = 0.98$$

b) Varying depth

$$\frac{\overline{X}}{W} = 0.5 \left(\frac{X_{in}}{W} + \frac{X_{out}}{W} \right) = 0.5 \cdot (0.96 \cdot 0.953) = 0.9565$$

$$\frac{\overline{H}}{H_1} = 0.5 \left(\frac{H_{in}}{H_1} + \frac{H_{out}}{H_1} \right) = 0.5 \cdot \left(\frac{9}{9} + \frac{8.625}{9} \right) = 0.9792$$

$$\frac{G_s}{G} = \frac{\overline{X}\,\overline{H}}{W\,H_1} = 0.9366$$

The profiles of stock temperature and pressure can be calculated from the melting profile by using the width of the melt filled part of the channel in the equations as shown in the following Section.

4.2.3 Melt Temperature

The exact calculation of melt or stock temperature can be done only on an iterative basis as shown in the computer program TEMPMELT listed in the Appendix. The following relationships and the numerical example serve as good estimates of the stock temperature.

Temperature rise ΔT:

$$\Delta T = \left(T_{\text{out}} - T_{\text{M}}\right) = \frac{3600\left(Z_{\text{c}} + Z_{\text{FLT}} + N_{\text{H}}\right)}{\dot{m}\cdot c_{\text{pm}}} \tag{4.31}$$

Heat through the barrel or heat lost from the melt:

$$N_{\text{H}} = \frac{\alpha_{\text{sz}}\cdot\pi\cdot D_{\text{FLT}}\cdot\Delta L\left(T_{\text{b}} - T_{\text{EIN}}\right)}{10^{6}\cdot\cos\phi} \tag{4.32}$$

Example for calculating N_{H} with symbols and units

$\alpha_{\text{s}} = 315.5$ W/(m² K); $D_{\text{FLT}} = 59$ mm; $\Delta L = 600$ mm; $T_{\text{b}} = 150$ °C; $c_{\text{pm}} = 2$ kJ/(kg·K)

Stock temperature at the inlet of the screw increment considered:

$T_{\text{in}} = 200$ °C

N_{H} from Eq. (4.32)

$$N_{\text{H}} = \frac{315.5\cdot\pi\cdot59.8\cdot600\cdot50}{10^{6}\cdot\cos17.66°}$$

ΔT with the values $Z_{\text{c}} = 3.84$ kW, $Z_{\text{FLT}} = 1.56$ kW, and $\dot{m} = 41.8$ kg/h from the earlier example and from Eq. (4.31)

$$\Delta T = \frac{3600\cdot3.54}{41.8\cdot2} = 152.4\,°C$$

Stock temperature at the outlet of the screw increment considered T_{out}:

$$T_{\text{out}} = T_{\text{M}} + 152.4\,°C$$

Melting point of the polymer $T_{\text{M}} = 110$ °C

Hence $T_{\text{out}} = 110 + 152.4 = 262.4$ °C

Average stock temperature $\bar{T}$:

$$\bar{T} = \frac{T_{\text{in}} + T_{\text{out}}}{2} = \frac{200 + 262.4}{2} = 231.2\,°C$$

As already mentioned, this result can only be an estimate because the effect of the change of temperature on the viscosity can be calculated only through an iterative procedure, as shown in [13].

4.2.4 Melt Pressure

For a screw zone of constant depth the melt or stock pressure can be obtained generally from the pressure flow by means of Eq. (4.3). However, the following empirical equation [13] has been found to give good results in practice:

$$|\Delta p| = \frac{F_1 \cdot 2 \cdot \eta_p \cdot \Delta l}{\sin\phi \left(H_{out} + \delta_{FLT}\right)} \cdot \left[\frac{\left|\dot{Q}_p\right| \left(2\,\eta_R + 1\right) \cdot \left(H_R + H_R^2\right) \cdot 10^9}{W\left(H_{out} + \delta_f\right)^2 \cdot n_R \cdot v} \right]^{n_R} \cdot 10^{-5} \tag{4.33}$$

where

$$\eta_p = \frac{\eta_\alpha}{\dot{\gamma}^{n_R - 1}} \tag{4.34}$$

The sign of Δp corresponds to that of the pressure flow $\dot{Q}_p$.

Example with Symbols and Units

a) Screw zone of constant channel depth (metering zone)

Empirical factor	$F_1 = 0.286$
Melt viscosity in screw channel	$\eta_\alpha = 1400$ Pa·s
Shear rate in channel	$\dot{\gamma} = 84$ s^{-1}
Length of screw zone (or of an increment)	$\Delta l = 600$ mm
Helix angle	$\phi = 17.66°$
Channel depth at the outlet of the zone or increment	$H_{out} = 3$ mm
Flight clearance	$\delta_{FLT} = 0.1$ mm
Pressure flow	$\dot{Q}_p = 1.249 \cdot 10^{-6}$ m^3/s
Reciprocal of the power law exponent n	$n_R = 0.5$
Ratio of channel depths at the outlet (H_{out}) and inlet (H_{in}) of the zone or increment H_R	$H_R = 1$ (constant depth)
Width of the channel	$W = 51.46$ mm
Thickness of the melt film	$\delta_f = 0$
Number of flights	$v = 1$

η_p from Eq. (4.34):

$$\eta_p = \frac{1400}{84^{0.5-1}} = 12831$$

Δp from Eq. (4.33):

$$\Delta p = \left[\frac{0.286 \cdot 2 \cdot 600 \cdot 12831}{\sin 17.66° (3 + 0.1)}\right] \cdot \left[1.249 \cdot 10^{-6} (2 \cdot 0.5 + 1) \cdot \frac{(1+1) \cdot 10^9}{51.46 \cdot 3^2 \cdot 0.5 \cdot 1}\right]^{0.5} \cdot 10^{-5}$$

$$= 218 \text{ bar}$$

b) Screw zone of varying depth (transition zone)

$H_{in} = 9$ mm; $H_{out} = 3$ mm; $\Delta l = 240$ mm; $\eta = 1800$ Pa·s; $\dot{\gamma} = 42$ s^{-1}

η_p from Eq. (4.34):

$$\eta_p = \frac{1800}{42^{0.5-1}} = 11665$$

Δp from Eq. (4.33):

$$\Delta p = \left[\frac{0.286 \cdot 2 \cdot 240 \cdot 11665}{\sin 17.66° (3 + 0.1)}\right] \cdot \left\{1.249 \cdot 10^{-6} \cdot \left[\left(\frac{3}{9}\right) + \left(\frac{3}{9}\right)^2\right] \cdot \frac{10^9}{51.46 \cdot 3^2 \cdot 0.5}\right\}^{0.5} \cdot 10^{-5}$$

$$= 37.2 \text{ bar}$$

A more exact calculation of the melt pressure profile in an extruder should consider the effect of the ratio of pressure flow to drag flow, the so called drossel quotient, as shown in [13].

4.2.5 Heat Transfer between the Melt and the Barrel

To estimate the power required to heat the barrel or to calculate the heat lost from the melt, the heat transfer coefficient of the melt at the barrel wall is needed. This can be estimated from [14, 23]

$$\alpha_{sz} = \lambda_m \left(\frac{N}{60 \cdot \pi \cdot a}\right)^{0.5} \left\{1 - \frac{(T_f - T_m)[1 - \exp(\beta)]}{(T_b - T_m)}\right\} \qquad (4.35)$$

where the thermal diffusivity a

$$a = \frac{\lambda_m}{10^6 \cdot c_m \cdot \rho_m} \qquad (4.36)$$

and the parameter β

$$\beta = -\frac{10^{-6} \cdot \delta_{FLT}^2 \cdot N}{240 \cdot a} \qquad (4.37)$$

Indices:

m: melt

f: melt film

b: barrel

Calculated Example with Symbols and Units

Thermal conductivity $\lambda_m = 0.174$ W/(m·K)

Specific heat $c_{pm} = 2$ kJ/(kg·K)

Melt density $\rho_m = 0.7$ g/cm^3

Thermal diffusivity a from Eq. (4.36)

$$a = 1.243 \cdot 10^{-7} \text{ m}^2/\text{s}$$

Flight clearance $\delta_{FLT} = 0.1$ mm

Screw speed $N = 80$ rpm

Parameter β from Eq. (4.37):

$$\beta = 0.027$$

For $T_f = 137.74$ °C, $T_m = 110$ °C and $T_b = 150$ °C

α_{sz} from Eq. (4.35):

$$\alpha_{sz} = 0.174 \left(\frac{80 \cdot 10^7}{60 \cdot \pi \cdot 1.243} \right)^{0.5} \left\{ 1 - \frac{(137.7 - 110)\left[1 - \exp(-0.027) \right]}{150 - 110} \right\} = 315.5 \text{ W/(m}^2 \cdot \text{K)}$$

4.2.6 Screw Power

The screw power consists of the power dissipated as viscous heat in the channel and flight clearance and the power required to raise the pressure of the melt. The total power Z_N for a melt filled zone is therefore [13]

$$Z_N = Z_c + Z_{FLT} + Z_{\Delta p}$$ (4.38)

where

Z_c = power dissipated in the screw channel

Z_{FLT} = power dissipated in the flight clearance

$Z_{\Delta p}$ = power required to raise the pressure of the melt

The power dissipated in the screw channel Z_c is given by [14]

$$Z_c = \frac{v \cdot \pi^2 \cdot D_{FLT}^2 \cdot N^2 \cdot W \cdot \eta_c \cdot \Delta L \left(F_z \cos^2 \phi + 4 \sin^2 \phi \right)}{36 \cdot 10^{14} \cdot \delta_{FLT} \cdot \sin \phi}$$ (4.39)

The power dissipated in the flight clearance can be calculated from [13]

$$Z_{FLT} = \frac{v \cdot \pi^2 \cdot D_{FLT}^2 \cdot N^2 \cdot w_{FLT} \cdot \eta_{FLT} \cdot \Delta L}{36 \cdot 10^{14} \cdot \delta_{FLT} \cdot \sin \phi}$$ (4.40)

The power required to raise the pressure of the melt $Z_{\Delta p}$ can be written as

$$Z_{\Delta p} = 100 \cdot \dot{Q}_p \cdot \Delta p$$ (4.41)

The flight diameter D_{FLT} is obtained from

$$D_{FLT} = D_b - 2 \cdot \delta_{FLT}$$ (4.42)

and the channel width W

$$W = \frac{s}{v} \cos \phi - w_{FLT}$$ (4.43)

The symbols and units used in the equations above are given in the following example.

Calculated Example

For the following conditions the screw power is to be determined:

Resin: LDPE with the constants of viscosity

$A_0 = 4.2968$

$A_1 = -3.4709 \cdot 10^{-1}$

$A_2 = -1.1008 \cdot 10^{-1}$

$A_3 = 1.4812 \cdot 10^{-2}$

$A_4 = -1.1150 \cdot 10^{-3}$

$b_1 = 1.29 \cdot 10^5$

$b_2 = 4.86 \cdot 10^3$ K

Melt density $\rho_m = 0.7$ g/cm^3

Operating conditions:

Screw speed $N = 80$ rpm

Melt temperature $T = 200$ °C

Die pressure $\Delta p = 300$ bar

Geometry of the melt zone or metering zone:

$D = 60$ mm; $H = 3$ mm; $e = 6$ mm; $s = 60$ mm; $\delta_{FLT} = 0.1$ mm; $\Delta L = 600$ mm; $v = 1$

Solution:

Power Z_c in the screw channel:

$D_{FLT} = 59.8$ mm from Eq. (4.42)

Shear rate in the screw channel $\dot{\gamma}_c$:

$\gamma_c = 83.8$ s 1 from Eq. (4.40)

$a_T = 0.374$

Viscosity of the melt in the screw channel η_c:

$\eta_c = 1406.34$ Pa·s

Channel width W:

$W = 51.46$ mm from Eq. (4.43)

Number of flights v:

$v = 1$

Length of the melt zone ΔL:

$\Delta L = 600$ mm

Factor F_x:

$F_x = 1$ for $\dfrac{H}{W} = \dfrac{3}{51.46} = 0.058$ from Eq. (4.14)

Helix angle ϕ:

$\phi = 17.66°$; $\sin \phi = 0.303$ from Eq. (4.2)

Power in the screw channel Z_c from Eq. (4.39)

$$Z_c = 1 \cdot \pi^2 \cdot 59.8^2 \cdot 80^2 \cdot 51.46 \cdot 1406.34 \cdot 600 \cdot \frac{\left(1 \cdot \cos^2 17.66° + 4 \sin^2 17.66°\right)}{36 \cdot 10^{14} \cdot 3 \cdot \sin 17.66°} = 3.84 \text{ kW}$$

Power in the flight clearance Z_{FLT}:

Flight width w_{FLT}

$$w_{FLT} = e \cos\phi = 6 \cdot \cos 17.66° = 5.7 \text{ mm}$$

Shear rate in the flight clearance $\dot{\gamma}_{FLT}$:

$$\dot{\gamma}_{FLT} = \frac{\pi \cdot D_b \cdot N}{60 \cdot \delta_{FLT}} = \frac{\pi \cdot 60 \cdot 80}{60 \cdot 0.1} = 2513.3 \text{ s}^{-1}$$

Shift factor a_T:

$a_T = 0.374$

Viscosity in the flight clearance η_{FLT}:

$\eta_{FLT} = 219.7$ Pa·s

Length of the melt zone ΔL:

$\Delta L = 600$ mm

Z_{FLT} from Eq. (4.40)

$$Z_{FLT} = \frac{1 \cdot \pi^2 \cdot 59.8^2 \cdot 80^2 \cdot 600 \cdot 5.7 \cdot 219.7}{36 \cdot 10^{14} \cdot 0.1 \cdot 0.303} = 1.56 \text{ kW}$$

Power to raise the melt pressure $Z_{\Delta p}$

Pressure flow $\dot{Q}_p$:

$\dot{Q}_p$ from the Example in Section 4.1.1.2

$$\dot{Q}_p = 1.249 \cdot 10^{-6} \text{ m}^3/\text{s}$$

Die pressure Δp:

$\Delta p = 300$ bar

$Z_{\Delta p}$ from Eq. (4.41)

$$Z_{\Delta p} = 100 \cdot 1.249 \cdot 10^{-6} \cdot 300 = 0.0375 \text{ kW}$$

Hence the power $Z_{\Delta p}$ is negligible in comparison with the sum $Z_c + Z_{FLT}$.

4.2.7 Temperature Fluctuation of the Melt

Temperature and pressure variations of the melt in an extruder serve as a measure for the quality of the extrudate and provide information as to the performance of the screw.

The temperature variation ΔT may be estimated from the following empirical relation, which was developed from the results of experiments of Squires [15] conducted on 3-zone screws:

$$\Delta T = \frac{5}{9}\left(\frac{1}{4.31\,N_Q^2 - 0.024}\right) \tag{4.44}$$

This relation is valid for $0.11 < N_Q < 0.5$.

The parameter N_Q is given by

$$N_Q = 14.7 \cdot 10^{-4}\, \frac{D_b^2}{G} \sum \frac{L}{H} \tag{4.45}$$

where

$\Delta T =$ temperature variation (°C)

$D_b =$ barrel diameter (cm)

$G \;\;=$ extruder output (g/s)

$L \;\;=$ length of screw zone in diameters

$H \;\;=$ depth of the screw zone (cm)

Calculated Example

The following values are given:

D_b = 6 cm
G = 15 g/s

L	depth cm	L/H
9	0.9	10
3	0.6 (mean value)	5
9	0.3	30

Hence $\sum \dfrac{L}{H} = 45$

N_Q from Eq. (4.45): $N_Q = 14.7 \cdot 10^{-4} \cdot \dfrac{36}{15} \cdot 45 = 0.159$

ΔT from Eq. (4.44): $\Delta T = \dfrac{5}{9}\left(\dfrac{1}{4.31 \cdot 0.159^2 - 0.024}\right) = 6.57\,°C \pm 3.29\,°C$

The constants occurring in the Eq. (4.44) and Eq. (4.45) depend on the type of polymer used. For screws other than 3-zone screws the geometry term in Eq. (4.45) has to be defined in such a way that N_Q correlates well with the measured temperature fluctuations.

4.2.8 Pressure Fluctuation

The effect of pressure fluctuations of the melt on the flow rate can be estimated by the relationship [4]

$$\dot{Q}_1 / \dot{Q}_2 = \left(\Delta p_1 / \Delta p_2 \right)^n \tag{4.46}$$

Calculated Example

Pressure drop $\Delta p_1 = 200$ bar; $\Delta p_2 = 210$ bar

Flow rate $\dot{Q}_1 = 39.3$ cm^3/s

Power law exponent n for LDPE = 2.5

Calculate the flow rate variation.

Solution:

$$\dot{Q}_1 / \dot{Q}_2 = 39.7 / \dot{Q}_2 = \left(200 / 210 \right)^{2.5} = 0.885$$

$$\dot{Q}_2 = 44.9 \text{ cm}^3/\text{s}$$

Flow rate variation $= \left(\dot{Q}_2 - \dot{Q}_1 \right) / \dot{Q}_1 = \dfrac{44.9 - 39.7}{39.7} = 13\%$

Pressure variation $= \dfrac{\Delta p_2 - \Delta p_1}{\Delta p_1} = \dfrac{210 - 200}{200} = 5\%$

This means that a pressure variation of 5% can cause a flow rate fluctuation of 13%. ∎

4.2.9 Extrusion Screw Simulations

The performance of an extrusion screw employed in a process can be simulated by using the equations for the melting profile, stock temperature, and melt pressure given in the Section 4.2 and, if necessary, the screw design optimized for better performance by using the program TEMPMELT mentioned in the Appendix. The results of a simulation are shown schematically in Fig. 4.24 [17]. The following practical examples illustrate this procedure.

Example 1: Poor Melt Quality in an Extrusion Coating Process

Figure 4.3 showed the melting of a polymer in the screw channels of an extrusion screw. The melt quality can be defined by the ratio G_s/G, as depicted in Fig. 4.24. The melting is completed, if this ratio is zero [1].

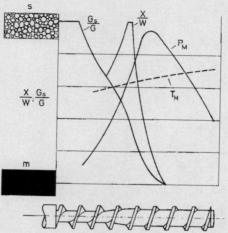

FIGURE 4.24 Schematic screw simulation profiles.
G_s: solids flow rate, G: Total flow rate, X: Width of solids bed, W: Channel width, P_M: Melt Pressure, T_M: Melt temperature

This example refers to an extrusion problem [9] concerning melt quality, which a processor had while extruding LDPE in a coating process. Computer simulations of the screw used by the processor at two different output rates showed that the solids content of the polymer at the entrance of the mixing section, particularly at the higher output, was too high to be melted and homogenized by the mixing device as shown in Fig. 4.25.

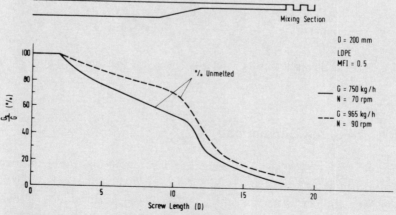

FIGURE 4.25 Poor melt quality of a processor's screw

The enlarged melting profiles in the metering zone presented in Fig. 4.26 show clearly that the inhomogeneous melt results from the improper screw design and that the screw employed by the processor cannot produce a melt of good quality at the higher output desired by the processor.

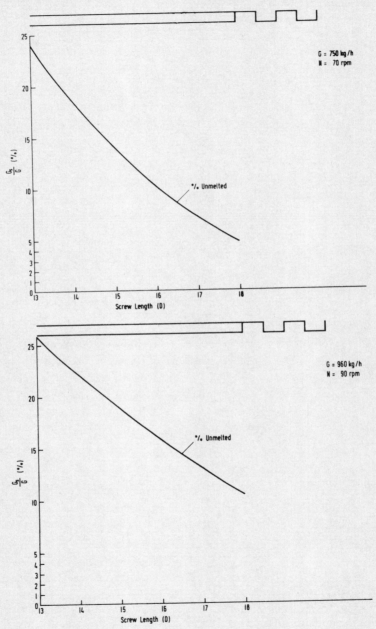

FIGURE 4.26 Enlarged melting profiles of the customer's screw at low and high outputs

The screw was redesigned by implementing different screw geometries with a mixing device (Fig. 4.29) into the program. As the enlarged melting profiles in the metering zone of this screw (given in the Fig. 4.27) show, the solids content at the inlet of the mixing device is now low enough to produce a homogeneous melt.

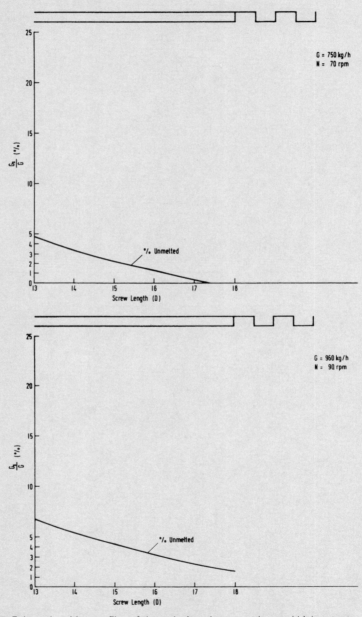

FIGURE 4.27 Enlarged melting profiles of the redesigned screw at low and high outputs

The processor obtained acceptable melt quality at the desired output by using the redesigned screw with shearing and distributive mixers, confirming the predictions of the program, as shown in Fig. 4.28.

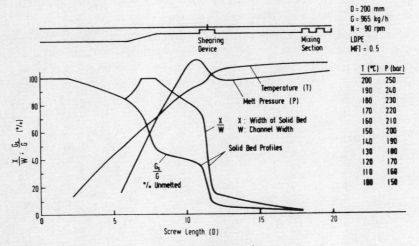

FIGURE 4.28 Redesigned screw with better melt homogeneity

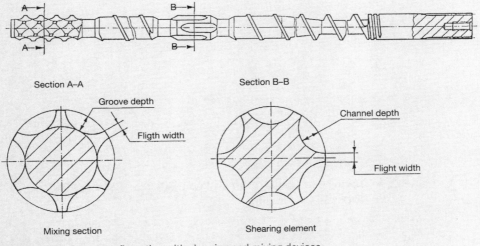

FIGURE 4.29 Screw configuration with shearing and mixing devices

Example 2: Determination of the Optimal Screw Geometry from a Given Number of Screws of Different Geometry Used in a Blown Film Process

The program can also be used to select a screw best suited for given operating conditions, among a set of existing screws of varied geometry.

A processor had five three-zone standard extrusion screws of different geometries, as shown in Fig. 4.30, at his disposal. The aim was to determine the screw geometry best suited for the operating conditions mentioned in Fig. 4.30. The calculated melting profiles in Fig. 4.30 show that screw 5 has the optimal geometry, because in this case complete melting has been attained. This predicted result was confirmed in practice.

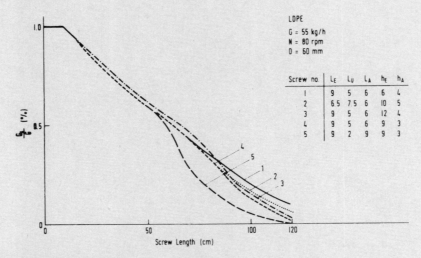

FIGURE 4.30 Determining optimal screw geometry

Example 3: Improving the Quality of Melt in an Extruder Used for Extrusion Coating and Laminating

This example deals with a screw used by a customer in a coating line for processing PET resin. It was found that the quality of the melt created by the screw was poor. As shown in Fig. 4.31, the simulation confirmed this observation of the customer.

The computer program also predicted that the quality of the melt can be improved by increasing the temperature of the pellets from 20 to 80 °C.

This measure was then employed in practice and led to a melt of better quality to the satisfaction of the customer (Fig. 4.31).

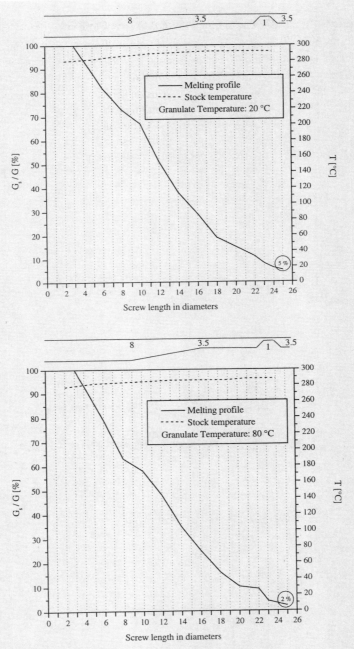

FIGURE 4.31 Melt quality at low and high granulate temperature

Example 4: Air Bubbles in a Flat Film Used in Packaging

Figure 4.32 shows air bubbles in a flat film of PA 6 exiting from a coat hanger die. By using different barrel temperature profiles, it was found that the temperature profile given in Fig. 4.33 with a high temperature near the feed zone would eliminate the bubbles. The predicted result was confirmed in practice.

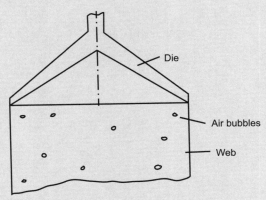

FIGURE 4.32 Air bubbles in a flat film of PA 6

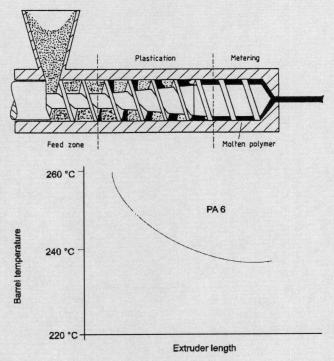

FIGURE 4.33 Barrel temperature used to eliminate air bubbles

The examples given above show that the use of computer programs in design work minimizes experimentation and in many cases eliminates it altogether.

4.2.10 Mechanical Design of Extrusion Screws

Torsion

The maximum shear stress τ_{max}, which occurs at the circumference of the screw root as a result of the torque M_T, is given by [10]

$$\tau_{max} = \frac{2 \cdot M_T}{\pi \cdot R^3} \tag{4.48}$$

where R = root radius of the screw.

The maximum feed depth H_{max} can be computed from [10]

$$H_{max} = 0.5 \cdot D - \left(\frac{2 \cdot M_T}{\pi \cdot \tau_{allow}}\right)^{\frac{1}{3}} \tag{4.49}$$

where

D = diameter

τ_{allow} = allowable shear stress of the screw metal

Calculated Example [10]

The maximum feed depth is to be calculated for the following conditions:

D = 150 mm; M_T = 17,810 Nm; τ_{allow} = 100 MPa;

H_{max} is found from Eq. (4.49):

$$H_{max} = \frac{1}{2} \cdot \frac{150}{1000} - \left(\frac{2 \cdot 17810}{\pi \cdot 100 \cdot 10^6}\right)^{\frac{1}{3}} = 0.075 - 0.0485 = 0.0265 \text{ m} \quad \text{or} \quad 26.5 \text{ mm}$$

Deflection

The lateral deflection of the screw caused by its own weight can be obtained from [10]

$$Y(L) = \frac{2 \cdot 1000 \cdot g \cdot \rho \cdot L^4}{E \cdot D^2} \tag{4.50}$$

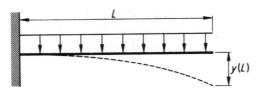

FIGURE 4.34 Lateral deflection of the screw as cantilever

Numerical Example with Symbols and Units [10]

g = 9.81 m^2/s   acceleration due to gravity

ρ = 7850 kg/m^3 density of the screw metal

L = 3 m   length of the screw

E = 210 · 10^9 Pa   elastic modulus of the screw metal

D = 0.15 m   screw diameter

Inserting these values into Eq. (4.50) one obtains

$$Y(L) = \frac{2 \cdot 1000 \cdot 9.81 \cdot 7850 \cdot 3^4}{210 \cdot 10^9 \cdot 0.15^2} = 2.64 \text{ mm}$$

This value exceeds the usual flight clearance, so that the melt between the screw and the barrel takes on the role of supporting the screw to prevent contact between the screw and the barrel [10].

Buckling Caused by Die Pressure

The critical die pressure, which can cause buckling, can be calculated from [10]

$$p_K = \frac{10^{-5} \cdot \pi^2 \cdot E}{64 \cdot \left(\dfrac{L}{D}\right)^2} \tag{4.51}$$

Numerical Example

E = 210 · 10^9 Pa   elastic modulus of the screw metal

L/D = 35   length to diameter ratio of screw

p_K from Eq. (4.51)

$$p_K = \frac{10^{-5} \cdot \pi^2 \cdot 210 \cdot 10^9}{64 \cdot 35^2} = 264.36 \text{ bar}$$

As can be seen from Eq. (4.50) the critical die pressure p_K decreases with increasing L/D ratio. This means that for the usual range of die pressures $(200 - 600$ bar) buckling through die pressure is a possibility, if the ratio L/D exceeds 20 [10].

Screw Vibration

When the screw speed corresponds to the natural frequency of lateral vibration of the shaft, the resulting resonance leads to large amplitudes, which can cause screw deflection. The critical screw speed according to [10] is given by

$$N_K = 0.88 \frac{D}{L^2}\left(\frac{E}{\rho}\right)^{\frac{1}{2}} \tag{4.52}$$

Substituting the values for steel, $E = 210 \cdot 10^9$ Pa and $\rho = 7850$ kg/m^3 we get

$$N_K = \frac{4549.5}{D \cdot \left(\dfrac{L}{D}\right)^2}$$

Calculated Example

For $D = 150$ mm and $L/D = 30$, NR is found from Eq. (4.52)

$$N_K = \frac{4549.5}{0.15 \cdot 30^2} = 33.7 \text{ rps} \quad \text{or} \quad 2022 \text{ rpm}$$

This result shows that for the normal range of screw speeds, vibrations caused by resonance are unlikely.

Uneven Distribution of Pressure

Non-uniform pressure distribution around the circumference of the screw can lead to vertical and horizontal forces of such magnitude that the screw deflects into the barrel. Even a pressure difference of 10 bar could create a horizontal force F_h in an extruder (diameter $D = 150$ mm within a section of length $L = 150$ mm)

$$F_h = \Delta p \cdot D \cdot L = 10 \cdot 10^5 \cdot 0.15^2 = 22.5 \text{ kN}$$

According to Rauwendaal [10], non-uniform pressure distribution is the most probable cause of screw deflection.

■ References

[1]   Rao, N. S.: Design Formulas for Plastics Engineers. Hanser, Munich (1991)

[2]   Rao, N. S.: Ruegg, T.: TAPPI PLC Conference Boston (2006)

[3]   N. N.: Brochure : Eastman 1997

[4]   Tadmor, Z.; Klein, I.: Engineering Principles of Plasticating Extrusion, Van Nostrand Reinhold, New York (1970)

[5]   Bernhardt, E. C.: Processing of Thermoplastic Materials, Van Nostrand Reinhold, New York (1963)

[6]   N. N.: Brochure: BASF AG 1992

[7]   Fritz, H. G.: Extrusion Blow Molding in Plastics Extrusion Technology, in F. Hensen (Ed.), Hanser, Munich (1988)

[8]   Darnell, W. H.; Mol, E. A. J.: *Soc. Plastics Eng. J.*, **12**, 20 (1956)

[9]   Rao, N. S.; O'Brien, K. T.; Harry, D. H.: Computer Modelling for Extrusion and Other Continuous Polymer Processes. In Keith T. O'Brien (Ed.), Hanser, Munich (1992)

[10]   Rauwendaal, C.: Polymer Extrusion 4th ed., Hanser, Munich (2001)

[11]   Rao, N.: Designing Machines and Dies for Polymer Processing with Computer Programs, Hanser, Munich (1981)

[12]   Rao, N.: Computer Aided Design of Plasticating Screws, Programs in Fortran and Basic, Hanser, Munich (1986)

[13]   Klein, L.; Marshall, D. I.: Computer Programs for Plastics Engineers, Reinhold, New York (1968)

[14]   Wood, S. D.: SPE 35, Antec (1977)

[15]   Squires, P. H.: *SPE J.*, **16** (1960), p. 267

[16]   Pearson, J. R. A.: Reports of University of Cambridge, Polymer Processing Research Centre (1969)

[17]   Rao, N.; Hagen, K.; Krämer, A.: *Kunststoffe*, **69** (1979) 10, p. 173

[18]   Schenkel, G.: Kunststoff-Extrudiertechnik. Hanser, Munich (1963)

[19]   Fenner, R. T.: Extrusion Screw Design, Illife Books, London (1970)

[20]   Fischer, P.: Dissertation, RWTH Aachen (1976)

[21]   Potente, H. : Proceedings, 9 Kolloquium IKV Aachen (1978)

[22]   Muenstedt, H.: *Kunststoffe*, **68** (1978), p. 92

[23]   Kautz, G.; Rao, N.: *Kunststoffe*, **66** (1976), p. 9

5 Analytical Procedures for Troubleshooting Extrusion Dies

The design of extrusion dies is based on the principles of rheology, thermodynamics, and heat transfer, which have been dealt with in Chapters 1 to 3. The quantities to be calculated are pressure, shear rate, and residence time as functions of the flow path of the melt in the die. The pressure drop is required to predict the performance of the screw and the die. Knowledge of shear rates in the die provides information on whether the melt flows within the range of permissible shear rates. Undue heating of the melt can be avoided on the basis of information on the residence time of the melt in the die, which also gives an indication of the uniformity of the melt flow.

The interaction between screw and die is shown in Fig. 5.1.

Figure 5.1 shows that the die temperature, die opening and channel depth significantly influence the extruder output.

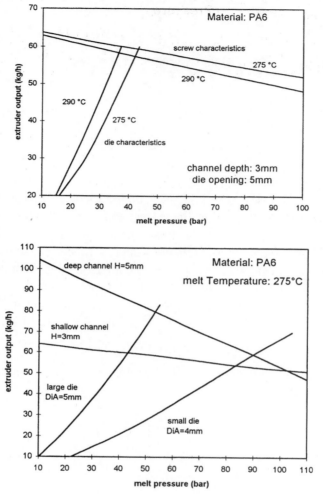

FIGURE 5.1 Interaction between screw and die

■ 5.1 Calculation of Pressure Drop

From the relationship between the volume flow rate and pressure drop of the melt in a die given by Eq. (1.10), it follows

$$\Delta p = \frac{\dot{Q}^{\frac{1}{n}}}{K^{\frac{1}{n}} \, G}$$

(5.1)

5.1.1 Effect of Die Geometry on Pressure Drop

The die constant G depends on the geometry of the die. Circle, slit, and annular cross-sections represent the flow channels of extrusion dies in most cases. G for a slit die is given by

$$G_{\text{slit}} = \left(\frac{W}{6}\right)^{\frac{1}{n}} \cdot \frac{H^{\frac{2}{n}+1}}{2L}$$

(5.2)

for $\dfrac{W}{H} \geq 20$

where H is the height of the slit and W is the width.

For $\dfrac{W}{H} < 20$, G_{slit} has to be multiplied by the correction factor F_{p} given in Fig. 5.2.

The factor F_{p} can be expressed as

$$F_{\text{p}} = 1.008 - 0.7474 \cdot \left(\frac{H}{W}\right) + 0.1638 \left(\frac{H}{W}\right)^2$$

(5.3)

The die constant G_{annulus} is calculated from

$$H = R_{\text{o}} - R_{\text{i}}$$

(5.4)

and

$$W = \pi \left(R_{\text{o}} + R_{\text{i}}\right)$$

(5.5)

where R_{o} is the outer radius and R_{i} is the inner radius. G_{annulus} then follows from Eq. (5.2)

$$G_{\text{annulus}} = \left(\frac{\pi}{6}\right)^{\frac{1}{n}} \frac{\left(R_{\text{o}} + R_{\text{i}}\right)^{\frac{1}{n}} \cdot \left(R_{\text{o}} - R_{\text{i}}\right)^{\frac{2}{n}+1}}{2L}$$

(5.6)

for values of the ratio $\pi \left(R_{\text{o}} + R_{\text{i}}\right) / \left(R_{\text{o}} - R_{\text{i}}\right) \geq 37$

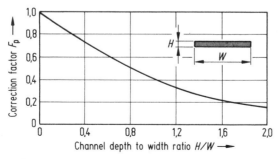

FIGURE 5.2 Correction factor F_{p} as a function of H/W [5]

For smaller values of this ratio, $G_{annulus}$ has to be multiplied by the factor F_p given in Fig. 5.2. The height H and width W are obtained in this case from Eq. (5.4) and Eq. (5.5).

5.1.2 Shear Rate in Die Channels

The shear rate for the different shapes of the channels treated above can be computed from [2]

$$\dot{\gamma}_{a_{circle}} = \frac{4\,\dot{Q}}{\pi\,R^3} \tag{5.7}$$

$$\dot{\gamma}_{a_{slit}} = \frac{6\cdot\dot{Q}}{W\cdot H^2} \tag{5.8}$$

$$\dot{\gamma}_{a_{annulus}} = \frac{6\cdot\dot{Q}}{\pi\left(R_o + R_i\right)\left(R_o - R_i\right)^2} \tag{5.9}$$

The shear rate for an equilateral triangle is given by [3]

$$\dot{\gamma}_{a_{angle}} = \frac{10}{3}\cdot\frac{\dot{Q}}{d^3} \tag{5.10}$$

where d is the half length of a side of the triangle.

The relationship for a square cross-section is found to be [3]

$$\dot{\gamma}_{a_{square}} = \frac{3}{0.42}\cdot\frac{\dot{Q}}{a^3} \tag{5.11}$$

where a is the length of a side of the square.

In the case of channels with varying cross-sections along the die length, for example, convergent or divergent channels, the channel is divided into a number of sufficiently small increments and the average dimensions of each increment are used in the equations given above [2].

5.1.3 General Relationship for Pressure Drop in Any Given Channel Geometry

By means of the equivalent radius defined by Schenkel [4] the pressure drop in cross-sections other than the ones treated in the preceding sections can be calculated. The equivalent radius is expressed by [4]

$$R_{\text{rh}} = \left[\frac{\left(\frac{2^{n+1}}{\pi} \right) \cdot A^{n+2}}{B^{n+1}} \right]^{\frac{1}{n+3}}$$

(5.12)

where

R_{rh} = equivalent radius

A = cross-sectional area

B = circumference

Another method of calculating the pressure drop in channels of varied cross-section is presented by Michaeli [24]. The correction factor F_p and the flow coefficient f_p [24] have the same values in comparable ranges of height to width of the channel.

5.1.4 Examples for Calculating Pressure Drop in the Die Channels of Different Shapes

Example 1: Pressure Drop in a Slit

This example shows how pressure can be calculated in a flat die in a cross-section with the shape of a slit.

Melt flows through a slit of width $W = 75$ mm, height $H = 1$ mm and length $L = 100$ mm.

The resin is LDPE with the viscosity constants given below. The mass flow rate and the melt temperature have the same values, $\dot{m} = 10$ g/s and $T = 200$ °C. The pressure drop Δp is to be calculated.

The constants of viscosity for the given LDPE [33] are

$A_0 = 4.2968$

$A_1 = -3.4709 \cdot 10^{-1}$

$A_2 = -1.1008 \cdot 10^{-1}$

$A_3 = 1.4812 \cdot 10^{-2}$

$A_4 = -1.1150 \cdot 10^{-3}$

$b_1 = 1.29 \cdot 10^5$

$b_2 = 4.86 \cdot 10^3$ K

Melt density $\rho_m = 0.7$ g/cm^3

Solution:

Volumetric flow rate $\dot{Q} = \dfrac{\dot{m}}{\rho_m} = \dfrac{0.01}{700} = 1.429 \cdot 10^{-5} \ \text{m}^3/\text{s}$

Shear rate:

$$\dot{\gamma}_a = \frac{6 \cdot \dot{Q}}{W \cdot H^2} = \frac{6 \cdot 1.429 \cdot 10^{-5}}{0.075 \cdot 0.001^2} = 1143.2 \ \text{s}^{-1}$$

Shift factor a_T:

$$a_T = b_1 \cdot \exp\left(b_2 / T\right) = 1.29 \cdot 10^5 \exp\left[4860 / \left(200 + 273\right)\right] = 0.374 \tag{5.13}$$

Power law exponent n from Eq. (5.14)

$$\frac{1}{n} = 1 + A_1 + 2 \, A_2 \, \lg\left(a_T \cdot \dot{\gamma}_a\right) + 3 \cdot A_3 \left[\lg\left(a_T \cdot \dot{\gamma}_a\right)\right]^2 + 4 \, A_4 \left[\lg\left(a_T \cdot \dot{\gamma}_a\right)\right]^3 \tag{5.14}$$

$n = 3.334$

Viscosity η_a from Eq. (5.15) [33]

$$\begin{aligned} \lg \eta_a = {} & \lg a_3 + A_0 + A_1 \, \lg\left(a_T \cdot \dot{\gamma}_a\right) + A_2 \left[\lg\left(a_T \cdot \dot{\gamma}_a\right)\right]^2 \\ & + A_3 \left[\lg\left(a_T \cdot \dot{\gamma}_a\right)\right]^3 + A_4 \left[\lg\left(a_T \cdot \dot{\gamma}_a\right)\right]^4 \end{aligned} \tag{5.15}$$

with $a_T = 0.374$, $\dot{\gamma}_a = 1143.2 \ \text{s}^{-1}$ and the constants $A_0 \ldots A_4$ we obtain ($\lg = \log$ base 10)

$\eta_a = 257.6 \ \text{Pa} \cdot \text{s}$

Shear stress τ

$\tau = 294{,}524.9 \ \text{N/m}^2$

Constant of proportionality K:

$K = 6.669 \cdot 10^{-16}$

Correction factor F_p:

$\dfrac{W}{H} = 75$

As the ratio W/H is greater than 20, the die constant, which can be calculated from Eq. (5.2), need not be corrected.

$G_{\text{slit}} = 2.1295 \cdot 10^{-5}$

and finally the pressure drop, Δp, from Eq. (5.1) is

$\Delta p = 588.8 \ \text{bar}$

Example 2: Pressure Drop in an Annulus

Annulus occurs as a cross-section in a spider die treated in Section 5.2.

Melt flows through an annulus with an outside radius $R_0 = 40$ mm, an inside radius $R_i = 39$ mm and of length $L = 00$ mm.

The resin is LDPE with the same viscosity constants as in Example 1. The process parameters mass flow rate and melt temperature remain the same.

Solution:

Volume flow rate $\dot{Q} = \dfrac{\dot{m}}{\rho_m} = \dfrac{0.01}{700} = 1.429 \cdot 10^{-5} \text{ m}^3/\text{s}$

Shear rate $\dot{\gamma}_a$ from Eq. (5.9)

$$\dot{\gamma}_a = \frac{6\,\dot{Q}}{\pi \left(R_o + R_i\right)\left(R_o - R_i\right)^2} = \frac{6 \cdot 1.429 \cdot 10^{-5}}{\pi \left(0.04 + 0.039\right) 0.001^2} = 345.47 \text{ s}^{-1}$$

Shift factor a_T:

$a_T = 0.374$

Power law exponent n:

$n = 2.907$

Viscosity η_a:

$\eta_a = 579.43 \text{ Pa} \cdot \text{s}$

Shear stress τ:

$\tau = 200173.6 \text{ N/m}^2$

Constant of proportionality K:

$K = 1.34 \cdot 10^{-13}$

Correction factor F_p

As the ratio $\dfrac{\pi\left(R_o + R_i\right)}{R_o - R_i} = 248.19$ is greater than 37, no correction is necessary.

$G_{annulus}$ from Eq. (5.6):

$$G_{annulus} = \left(\frac{\pi}{6}\right)^{\frac{1}{n}} \cdot \left(R_o + R_i\right)^{\frac{1}{n}} \cdot \frac{\left(R_o - R_i\right)^{\frac{2}{n}+1}}{2L} = \frac{\left(\frac{\pi}{6}\right)^{\frac{1}{2.907}} \cdot \left(0.04 + 0.039\right)^{\frac{1}{2.907}} \cdot 0.001^{\frac{2}{2.907}+1}}{2 \cdot 0.1}$$

$$= 1.443 \cdot 10^{-5}$$

Pressure drop Δp from Eq. (5.1)

$$\Delta p = 400.26 \text{ bar}$$

Example 3: Pressure Drop in a Quadratic Cross-Section

Profile dies often have cross-sections shaped like squares or triangles. This example deals with the pressure drop in a square die.

Melt flows through a quadratic cross-section with the length of a side $a = 2.62$ mm. The channel length L is 50 mm. The resin is LDPE with the following constants occurring in the power law relationship

$$\eta_a = K_{OR} \exp(-\beta \cdot T) \cdot \dot{\gamma}^{n_R - 1} \tag{5.16}$$

K_{OR} = 135,990
β = 0.00863
n_R = 0.3286

Following conditions are given:

Mass flow rate $m = 0.01$ g/s
Melt temperature $T = 200$ °C
Melt density $\rho_m = 0.7$ g/cm^3

The pressure drop Δp is to be calculated.

Solution:

Three methods of calculation will be presented for this example.

Method a:

In this method the melt viscosity is calculated according to the power law. Otherwise, the method of calculation is identical to the one depicted in the foregoing examples.

Volume flow rate $\dot{Q} = \dfrac{\dot{m}}{\rho_m} = \dfrac{0.01}{0.7} = 0.014286 \text{ cm}^3/\text{s}$

Shear rate $\dot{\gamma}_a = \dfrac{6\,\dot{Q}}{W\,H^2}$

For a square with $W = H$ the shear rate $\dot{\gamma}_a$ is

$$\dot{\gamma}_a = \frac{6\,\dot{Q}}{H^3} = \frac{6 \cdot 0.014286}{0.262^3} = 4.76 \text{ s}^{-1}$$

Power law exponent n:

$$n = \frac{1}{n_R} = \frac{1}{0.3286} = 3.043$$

Viscosity η_a

$$\eta_a = K_R\,\exp(-\beta \cdot T) \cdot \dot{\gamma}^{n_R - 1} = 135990 \cdot \exp(-0.00863 \cdot 200) \cdot 4.766^{0.3286-1}$$
$$= 24205.54 \cdot 4.706^{0.3286-1} = 8484 \text{ Pa} \cdot \text{s}$$

Shear stress τ

$$\tau = 40434.88 \text{ N/m}^2$$

Constant of proportionality K

$$K = 4.568 \cdot 10^{-14}$$

Correction factor F_p

As $\dfrac{W}{H} = 1$ is less than 20, the correction factor is obtained

$$F_p = 1.008 - 0.7474 \cdot 1 + 0.1638 \cdot 1^2 = 0.4244$$

Die constant G_{slit}:

$$G_{slit} = 4.22 \cdot 10^{-5}$$

$$G_{slit\ corrected} = 0.4244 \cdot 4.22 \cdot 10^{-5} = 1.791 \cdot 10^{-5}$$

Pressure drop Δp:

$$\Delta p = 35.7 \text{ bar}$$

Method b:

The shear rate $\dot{\gamma}_a$ is calculated from Eq. (5.11) [3]

$$\dot{\gamma}_a = \frac{3}{0.42} \cdot \frac{\dot{Q}}{a^3} = \frac{3000 \cdot 0.014286}{0.42 \cdot 2.62^3} = 5.674 \ s^{-1}$$

Viscosity η_a

$$\eta_a = 24205.54 \cdot 5.674^{0.3286-1} = 7546.64 \ Pa \cdot s$$

Shear stress τ

$$\tau = 42819.6 \ N/m^2$$

Power law exponent n

$$n = \frac{1}{n_R} = 3.043$$

Proportionality factor K

$$K = 4.568 \cdot 10^{-14}$$

The pressure drop Δp is found from

$$\Delta p_{square} = \frac{2}{K^{\frac{1}{n}}} \cdot \left(\frac{3}{0.42}\right)^{\frac{1}{n}} \cdot \frac{\dot{Q}^{\frac{1}{n}}}{a^{\frac{3}{n}+1}} \cdot 2L \tag{5.17}$$

with the die constant G_{square}

$$G_{square} = \frac{1}{2}\left(\frac{0.42}{3}\right)^{\frac{1}{n}} \cdot \frac{a^{\frac{3}{n}+1}}{2L} = \frac{1}{2}\left(\frac{0.42}{3}\right)^{\frac{1}{3.043}} \cdot \left(\frac{2.62}{1000}\right)^{\frac{3}{3.043}+1} \cdot \frac{1}{0.05} = 1.956 \cdot 10^{-5} \tag{5.18}$$

In Eq. (5.17)

$$K^{\frac{1}{n}} = \left(4.568 \cdot 10^{-14}\right)^{\frac{1}{3.043}} = 4.134 \cdot 10^{-5}$$

$$\dot{Q}^{\frac{1}{n}} = \left(0.014286 \cdot 10^{-6}\right)^{\frac{1}{3.043}} = 2.643 \cdot 10^{-3}$$

Finally, Δp_{square} from Eq. (5.1):

$$\Delta p_{square} = \frac{\dot{Q}^{\frac{1}{n}}}{K^{\frac{1}{n}} \cdot G} = \frac{2.643 \cdot 10^{-3}}{4.134 \cdot 10^{-5} \cdot 1.956 \cdot 10^{-5}} = 32.686 \ bar$$

The above relationship for shear rate developed by Ramsteiner [3] leads therefore almost to the same result as the one obtained with Method a.

Method c:

In this method, first of all the equivalent radius is calculated from Eq. (5.12):

The pressure drop is then calculated using the same procedure as in the case of a round channel (Example 1):

Equivalent radius R_{rh}:

$$R_{rh} = \left[\frac{\frac{(2^{n+1})}{\pi} \cdot A^{n+2}}{B^{n+1}} \right]^{\frac{1}{n+3}}$$

$A = a^2 = 2.62^2 = 6.864 \text{ mm}^2$
$B = 4a = 4 \cdot 2.62 = 10.48 \text{ mm}$
$n = 3.043$
$R_{rh} = 1.363 \text{ mm}$

Shear rate $\dot{\gamma}_a$:

$$\dot{\gamma}_a = 7.183 \text{ s}^{-1}$$

Viscosity η_a

$$\eta_a = 24205.54 \cdot 7.183^{0.3216-1} = 6441.56 \text{ Pa} \cdot \text{s}$$

Shear stress τ

$$\tau = 46269.73 \text{ N/m}^2$$

Constant of proportionality

$$K = 4.568 \cdot 10^{-14}$$

G_{circle}

$$G_{circle} = \frac{\left(\frac{\pi}{4}\right)^{\frac{1}{3.043}} \cdot \left(\frac{1.363}{1000}\right)^{\frac{3}{3.043}+1}}{2 \cdot 0.05} = 1.885 \cdot 10^{-5}$$

Pressure drop from Eq. (5.1):

$$\Delta p_{square} = \frac{2.643 \cdot 10^{-3}}{4.134 \cdot 10^{-5} \cdot 1.885 \cdot 10^{-5}} = 33.92 \text{ bar}$$

This result shows that the relationship in Eq. (5.12) [4] is sufficiently accurate for practical purposes. This equation is particularly useful for dimensioning channels, whose geometries differ from the common shapes, namely, the circle, slit, or annulus. The procedure for calculating the pressure drop for an equilateral triangle is shown in the following example:

Example 4

Melt flows through an equilateral triangular channel of length $L = 50$ mm. The side of the triangle is $2 d = 4.06$ mm. Other conditions remain the same as in Example 3.

Solution:

Substitute radius R_{rh} from Eq. (5.12) with

$$A = \sqrt{3}\, d^2 = 7.1376 \text{ mm}^2$$
$$B = 3 \cdot 2\, d = 12.18 \text{ mm}$$
$$n = 3.043$$
$$R_{rh} = 1.274 \text{ mm}$$

Shear rate $\dot{\gamma}_a$ from Eq. (5.7):

$$\dot{\gamma}_a = \frac{4 \cdot 0.014286}{\pi \cdot 0.1274^3} = 8.8 \text{ s}^{-1}$$

Viscosity η_a

$$\eta_a = 24205.54 \cdot 8.8^{0.3286-1} = 5620.68 \text{ Pa} \cdot \text{s}$$

Shear stress τ

$$\tau = 49462 \text{ N/mm}^2$$

Constant of proportionality

$$K = 4.568 \cdot 10^{-14}$$

G_{circle}

$$G_{circle} = \frac{\left(\dfrac{\pi}{4}\right)^{\frac{1}{3.043}} \cdot \left(\dfrac{1.274}{1000}\right)^{\frac{3}{3.043}+1}}{2 \cdot 0.05} = 1.648 \cdot 10^{-5}$$

Pressure drop Δp from Eq. (5.1):

$$\Delta p = \frac{\dot{Q}^{\frac{1}{n}}}{K^{\frac{1}{n}} \cdot G_{circle}} = \frac{2.643 \cdot 10^{-3}}{4.134 \cdot 10^{-5} \cdot 1.648 \cdot 10^{-5}} = 38.79 \text{ bar}$$

Using the relation developed by Ramsteiner [3] on the basis of rheological measurement on triangular channels, Example 4 is calculated, as follows for the purpose of comparing both methods:

Shear rate:

$$\dot{\gamma}_a = \frac{10}{3} \cdot \frac{\dot{Q}}{d^3} = \frac{10}{3} \cdot \frac{0.014286}{2.03^3} = 5.692 \, s^{-1}$$

Viscosity η_a

$$\eta_a = 24205.54 \cdot 5.692^{0.3286-1} = 7530.61 \, Pa \cdot s$$

Shear stress τ

$$\tau = 42864.2 \, N/m^2$$

Constant of proportionality

$$K = 4.568 \cdot 10^{-14}$$

Die constant $G_{triangle}$:

$$G_{triangle} = \frac{1}{\sqrt{3}} \cdot \left(\frac{3}{10}\right)^{\frac{1}{n}} \cdot \frac{d^{\frac{3}{n}+1}}{2L} = \frac{1}{\sqrt{3}} \cdot \left(\frac{3}{10}\right)^{\frac{1}{3.043}} \cdot \frac{\left(\frac{2.03}{1000}\right)^{\frac{3}{3.043}+1}}{2 \cdot 0.05} = 1.75 \cdot 10^{-5} \qquad (5.19)$$

Pressure drop Δp from Eq. (5.1):

$$\Delta p = \frac{\dot{Q}^{\frac{1}{n}}}{K^{\frac{1}{n}} \cdot G_{triangle}} = \frac{2.643 \cdot 10^{-3}}{4.134 \cdot 10^{-5} \cdot 1.75 \cdot 10^{-5}} = 36.5 \, bar$$

This result differs little from the one obtained by using the concept of an equivalent radius. Therefore, the concept of Schenkel [4] is suitable for use in practice.

5.1.5 Temperature Rise and Residence Time

The adiabatic temperature increase of the melt can be calculated from

$$\Delta T = \frac{\Delta p}{10 \cdot \rho_m \cdot c_m} \quad (K) \qquad (5.20)$$

where

ΔT = temperature rise (K)

Δp = pressure difference (bar)

ρ_m = melt density (g/cm^3)

c_{pm} = specific heat of the melt (kJ/(kg·K))

The residence time $\bar{t}$ of the melt in the die of length L can be expressed as

$$\bar{t} = \frac{L}{\bar{u}}$$ (5.21)

$\bar{u}$ = average velocity of the melt

Equation (5.18) for a tube can be written as

$$t = \frac{4 \cdot L}{\dot{\gamma}_a \cdot R}$$ (5.22)

R = tube radius

$\dot{\gamma}_a$ = shear rate according to Eq. (5.7)

The relationship for the average resistance time in a slit is

$$\bar{t} = \frac{6 \cdot L}{\dot{\gamma}_a \cdot H}$$ (5.23)

H = height of slit

$\dot{\gamma}_a$ = shear rate according to Eq. (5.8)

■ 5.2 Spider Dies

Spider dies are used in blown film, pipe extrusion and blow molding processes. In general, the die consists of the geometrical sections given in Fig. 5.3.

For the input data shown in Table 5.1 the pressure drop, shear rate, and residence time along the die length for different process parameters are calculated using the equations for the round channel and annulus given in Section 5.1.4. For the spider cross-section the equation of Schenkel, Eq. (5.12), is used. The results of these calculations are shown in the Figures 5.4 to 5.6.

The aim here is to keep these variables within acceptable limits by changing the dimensions if necessary.

The computer program VISPIDER mentioned in the Appendix can be used for designing spider dies quickly and easily.

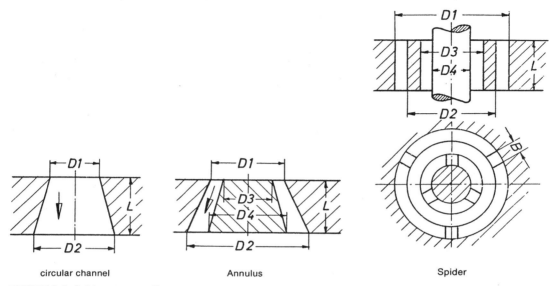

circular channel   Annulus   Spider

FIGURE 5.3 Spider cross-sections

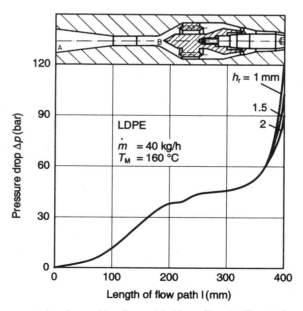

FIGURE 5.4 Calculated pressure drop in a spider die used in blown film at different die gaps

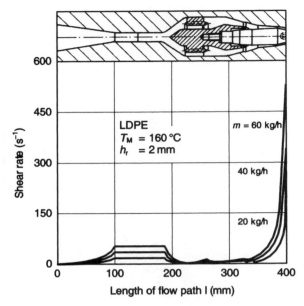

FIGURE 5.5 Shear rate along spider die

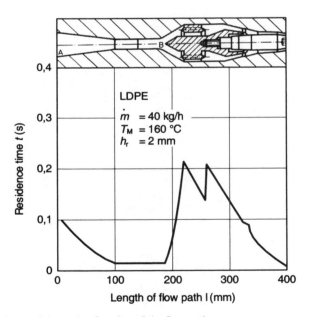

FIGURE 5.6 Residence time t of the melt a function of the flow path

TABLE 5.1 Pressure Drop Versus Spider Length

Flow Rate	30.00000 (kg/h)
D1:	45.00000 (mm)
D2:	16.00000 (mm)
L:	100.00000 (mm)
Pr. Drop:	9.65085 (bar)
D1:	16.00000 (mm)
D2:	16.00000 (mm)
L:	40.00000 (mm)
Pr. Drop:	11.02954 (bar)
D1:	16.00000 (mm)
D2:	16.00000 (mm)
L:	46.00000 (mm)
Pr. Drop:	12.68397 (bar)
D1:	16.00000 (mm)
D2:	68.00000 (mm)
D3:	0.0 (mm)
D4:	28.00000 (mm)
L:	36.00000 (mm)
Pr. Drop:	3.65644 (bar)
D1:	68.00000 (mm)
D2:	54.00000 (mm)
D3:	42.00000 (mm)
D4:	28.00000 (mm)
L:	33.00000 (mm)
N:	3.00000 (—)
B:	6.00000 (mm)
Pr. Drop:	13.47862 (bar)
D1:	68.00000 (mm)
D2:	68.00000 (mm)
D3:	28.00000 (mm)
D4:	44.00000 (mm)
L:	38.00000 (mm)
Pr. Drop:	1.48398 (bar)
D1:	68.00000 (mm)
D2:	54.00000 (mm)
D3:	44.00000 (mm)
D4:	35.00000 (mm)
L:	30.00000 (mm)
Pr. Drop:	2.62504 (bar)

TABLE 5.1   (continued)

Flow Rate	30.00000 (kg/h)
D1:	54.00000 (mm)
D2:	49.00000 (mm)
D3:	35.00000 (mm)
D4:	28.00000 (mm)
L:	8.00000 (mm)
Pr. Drop:	0.89935 (bar)
D1:	49.00000 (mm)
D2:	45.90599 (mm)
D3:	28.00000 (mm)
D4:	28.00000 (mm)
L:	5.50000 (mm)
Pr. Drop:	0.67009 (bar)
D1:	45.90599 (mm)
D2:	40.00000 (mm)
D3:	28.00000 (mm)
D4:	26.04500 (mm)
L:	10.50000 (mm)
Pr. Drop:	2.01417 (bar)
D1:	40.00000 (mm)
D2:	27.09999 (mm)
D3:	26.04500 (mm)
D4:	19.99500 (mm)
L:	32.50000 (mm)
Pr. Drop:	16.87939 (bar)
D1:	27.09999 (mm)
D2:	26.30600 (mm)
D3:	19.99500 (mm)
D4:	19.46599 (mm)
L:	1.50000 (mm)
Pr. Drop:	1.48500 (bar)
D1:	26.30600 (mm)
D2:	18.00000 (mm)
D3:	19.46599 (mm)
D4:	14.00000 (mm)
L:	15.50000 (mm)
Pr. Drop:	27.94635 (bar)
Total Pr. Drop:	104.50705 (bar)
Exit Temperature:	166.21030 (grad C)

■ 5.3 Spiral Dies

Spiral mandrel dies are used in blown film processes for making films with as small a deviation in the film thickness as possible. As shown in Fig. 5.7, the polymer melt is divided into several separate feed ports which empty into spirally shaped channels of decreasing cross-section cut into the mandrel. Between the die body and the land separating adjacent spirals there is a gap, or an annulus, into which a part of the spiral flow leaks. The flow proceeds upward over the land into the next spiral. It is the mixing of annular and spiral flows which leads to uniform melt flow distribution. A further advantage of spiral mandrel dies over spider dies is that weld lines and flow markings are eliminated due to the absence of mandrel support elements disturbing the flow.

FIGURE 5.7 Spiral die [1]

A planar view of a spiral die with four channels is shown in Fig. 5.8. Here the channels have a length which corresponds to only three quarters of the die circumference. The segment of the die which can be calculated by the computer program VISPIRAL listed in the Appendix, using the equations given in Section 5.1.4 is shown in Fig. 5.8. Due to reasons of symmetry this calculation is valid for other segments as well so that the whole die can be designed in this manner.

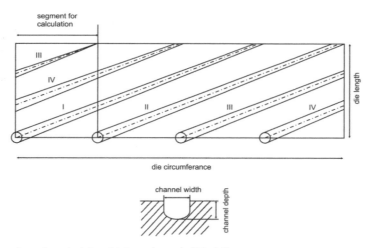

FIGURE 5.8 Planar view of a spiral die with four channels [33, 34]

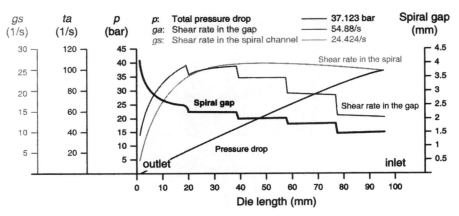

FIGURE 5.9 Results of simulation of a spiral die for LDPE with input data

The input data and calculated results of a die with three channels and a spiral length of one and two thirds of the die circumference are given in Fig. 5.9 for LDPE. The die gap required for uniform melt flow, the shear rates in the spiral and annulus as well as the pressure drop along the spiral are the output data needed to design the die.

■ 5.4 Adapting Die Design to Avoid Melt Fracture

Melt fracture can be defined as an instability of the melt flow leading to surface or volume distortions of the extrudate. Surface distortions [11] are usually created from instabilities near the die exit, while volume distortions [7] originate from the vortex formation at the die entrance. Due to the occurrence of these phenomena, melt fracture limits the production of articles manufactured by extrusion processes. The use of processing additives to increase the output has been dealt with in a number of publications given in [8]. However, processing aids are not desirable in applications such as pelletizing and blow molding. Therefore, the effect of die geometry on the onset of melt fracture has been examined.

Onset of Melt Fracture

The onset of melt fracture with increasing die pressure is shown for LDPE and HDPE in Fig. 5.10 [11]. As can be seen, the distortions appear differently depending on the resin. The volume flow rate is plotted in Fig. 5.11 [12], first as a function of wall shear stress, and then as a function of pressure drop in the capillary for LDPE and HDPE. The sudden increase in slope is evident for LDPE only in the plot flow rate vs. pressure, where as in the case of HDPE it is the opposite. Furthermore, for HDPE, the occurrence of melt fracture depends on the ratio length L to diameter D of the capillary. The effect of temperature on the onset of melt fracture is shown in Fig. 5.12 [9].

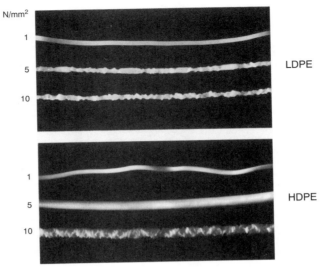

FIGURE 5.10 Irregularities of the extrudate observed at increasing extrusion pressure with LDPE and HDPE [11]

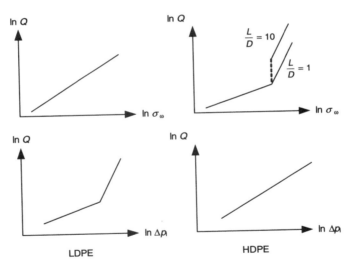

FIGURE 5.11 Volume flow rate vs. wall shear stress and vs. pressure drop in capillary for LDPE and HDPE [12]

With increasing temperature, the onset of instability shifts to higher shear rates. This behavior is used in practice to increase output. However, exceeding the optimum processing temperature can lead to a decrease in the quality of the product in many processing operations. From these considerations it can be seen that designing a die by taking the resin behavior into account is the easiest method to obtain quality products at high throughputs.

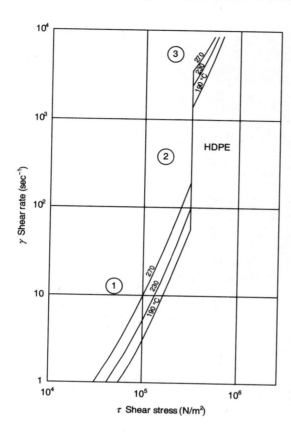

FIGURE 5.12 Effect of temperature on the melt fracture (region 2) for HDPE [9]

5.4.1 Pelletizer Dies

The aim here is to design a die for a given throughput or to calculate the maximum throughput possible without melt fracture for a given die. These targets can be achieved by performing simulations on dies of different tube diameters, flow rates, and melt temperatures. The program VISPIDER listed in the Appendix can be used for this case as well. Fig. 5.13 shows the results of one such simulation.

5.4.2 Blow Molding Dies

Figure 5.14 shows the surface distortion on the parison used in blow molding, which occurs at a definite shear rate depending on the resin. In order to obtain a smooth product surface, the die contour has been changed in such a way that the shear rate lies in a range that provides a smooth surface of the product (Fig. 5.15). The redesigned die creates lower extrusion pressures as well, as can be seen from Fig. 5.15 [9].

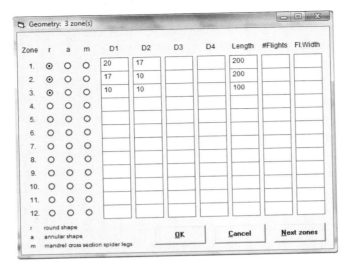

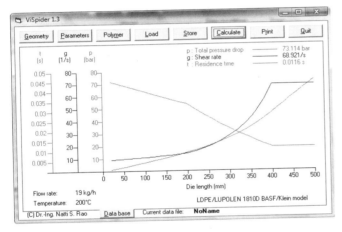

FIGURE 5.13 Results of simulation of a pelletizer die

FIGURE 5.14 Surface distortion on a parison used in the blow molding [9]

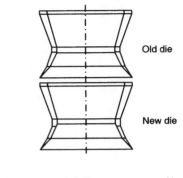

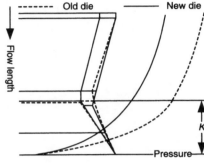

FIGURE 5.15 Die contour used for obtaining a smooth parison surface [9]

5.4.3 Summary of the Die Design Procedures

The following steps summarize the design procedures given above.

- STEP 1: Calculation of the shear rate in the die channel
- STEP 2: Fitting the measured viscosity plots with a rheological model
- STEP 3: Calculation of the power law exponent
- STEP 4: Calculation of the shear viscosity at the shear rate in Step 1
- STEP 5: Calculation of the wall shear stress
- STEP 6: Calculation of the factor of proportionality
- STEP 7: Calculation of die constant
- STEP 8: Calculation of pressure drop in the die channel and
- STEP 9: Calculation of the residence time of the melt in the channel

5.5 Flat Dies

Taking the resin behavior and the process conditions into account, the flat dies used in extrusion coating can be designed following similar guidelines as outlined above. Figure 5.16 shows the manifold radius required to attain uniform melt flow out of the die exit as a function of the manifold length [20].

The pressure drop in the flat film die can be obtained from the relationship [1]

$$\Delta p = \frac{2 \left(6 \cdot \dot{Q}\right)^{\frac{1}{n}} \cdot L}{K^{\frac{1}{n}} \cdot W^{\frac{1}{n}} \cdot H^{\frac{2}{n}+1}}$$

(5.24)

Calculated Example

Calculate Δp for the flat die geometry given in Fig. 5.16 for the conditions:

$\dot{m} = 36$ kg/h, $T = 200$ °C, $\rho_m = 0.7$ g/cm^3, $H = 2$ mm, $W = 75$ mm.

Viscosity coefficients for LDPE:

$a_0 = 3.388$, $a_1 = -0.635$, $a_{11} = -0.01815$, $a_2 = -0.005975$, $a_{22} = -0.0000025$, $a_{12} = 0.0005187$

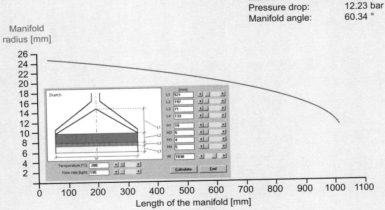

FIGURE 5.16 Manifold radius as a function of the distance along the length of the manifold

Solution:

Volume flow rate

$$\dot{Q} = \frac{\dot{m}}{\rho_m} = \frac{36}{3.6 \cdot 0.7} = 1.429 \cdot 10^{-5} \text{ m}^3/\text{s}$$

Shear rate

$$\dot{\gamma} = \frac{6\dot{Q}}{WH^2} = \frac{6 \cdot 1.429 \cdot 10^{-5}}{0.075 \cdot 0.002^2} = 285.7 \text{ s}^{-1}$$

Power law exponent:

$$n = 2.75$$

Viscosity:

$$\eta = 646.2 \text{ Pa} \cdot \text{s}$$

Shear stress:

$$\tau = 184619 \text{ N/m}^2$$

Proportionality constant:

$$K = 8.84 \cdot 10^{-13}$$

$$\Delta p = 184.62 \text{ bar}$$

Figures 5.17 and 5.18 show Δp calculated according to Eq. (5.24) as a function of the die gap H for the given die configuration and the polymers LDPE, LLDPE, and PET.

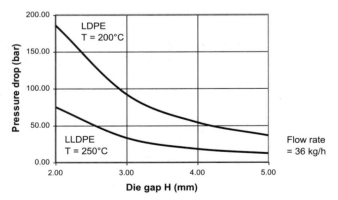

FIGURE 5.17 Pressure drop Δp as a function of die gap H for LDPE and LLDPE, $W = 75$ mm, $L = 100$ mm, $\dot{m} = 36$ kg/h

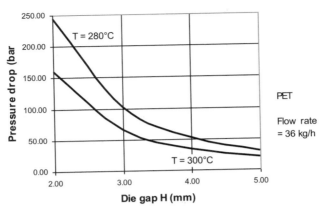

FIGURE 5.18 Effect of temperature on pressure drop Δp, $W = 75$ mm, $L = 100$ mm, $\dot{m} = 36\,\text{kg/h}$

Flat dies can be easily designed by using the program VISCOAT, listed in the Appendix. Programmable pocket calculators can handle the calculations as well.

5.6 An Easily Applicable Method of Designing Screen Packs for Extruders

In various extrusion processes, particularly those involving resin blends containing fillers and additives, it is often necessary to increase the melt pressure, in order to create more back mixing of the melt in the screw channel of the extruder. This can be achieved by using screen packs of different mesh sizes. They can also be used to increase the melt temperature to attain better plastication of the resin. Another application of screens concerns melt filtration, in which undesirable material is removed from the melt [19]. In all these operations it is necessary to be able to predict the pressure drop in the screen packs as accurately as possible, as the melt pressure is closely related to the extruder output. Based on recent developments in rheology this section presents an easy and quick method of calculating the pressure drop in a screen pack as a function of the resin viscosity, extruder throughput and the geometry of the screen. The effect of screen blocking is also taken into account. The predictions agree well with the experiments. Practical worked-out examples illustrate the design principles involved.

Mesh Size

The term screen as used here refers to a wire-gauze screen with a certain number of openings, depending on the size of the mesh. The mesh size is denoted by the mesh number which is the number of openings per linear inch counting from the center of any wire to a point exactly 25.4 mm or one inch distant [18] (Fig. 5.19). The mesh number m_n in Fig. 5.19, for instance, is 3. A screen pack consists usually of screens of varied mesh sizes with the screens placed one behind the other in a pack (Fig. 5.20).

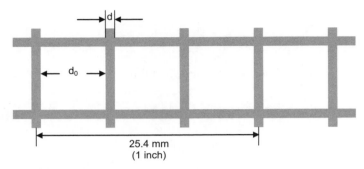

FIGURE 5.19 Mesh of a wire-gauze screen

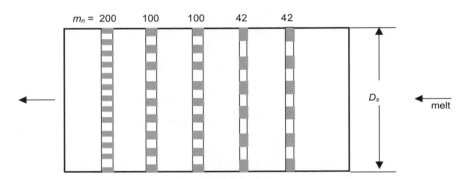

FIGURE 5.20 Screen pack with screens of varied mesh size

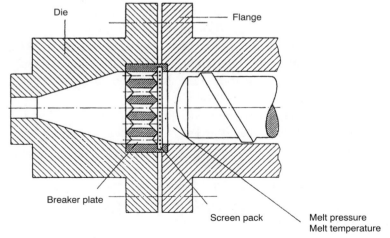

FIGURE 5.21 Position of screen pack in an extruder [17]

Design Procedure

The volume flow rate $\dot{q}$ through a hole for a square screen opening is given by [19]

$$\dot{q} = \frac{400 \cdot 6.45 \cdot \dot{M}}{3.6 \cdot \rho_m \cdot m_n^2 \cdot \pi \cdot D_s^2} \qquad (5.25)$$

The relationship between the volume flow rate $\dot{q}$ and the pressure drop Δp in the screen can be expressed as [6]

$$\dot{q} = K \cdot G^n \cdot \Delta p^n \qquad (5.26)$$

The shear rate of the melt flow for a square opening is calculated from [3]

$$\dot{\gamma}_{\text{square}} = \frac{3}{0.42} \cdot \frac{\dot{q}}{0.001 \cdot d_0^3} \qquad (5.27)$$

The power law exponent n follows from

$$\frac{1}{n} = 1 + a_1 + 2\,a_{11}\,\ln\dot{\gamma} + a_{12}\,T \qquad (5.28)$$

where a_1, a_{11} and a_{12} are viscosity coefficients in the Klein viscosity model and T is the melt temperature in °F.

Using the Klein model the melt viscosity η is obtained from

$$\ln\eta = a_0 + a_1\,\ln\dot{\gamma} + a_{11}\,\ln\dot{\gamma}^2 + a_2\,T + a_{22}\,T^2 + a_{12}\,T\,\ln\dot{\gamma} \qquad (5.29)$$

Shear stress τ:

$$\tau = \eta \cdot \dot{\gamma} \qquad (5.30)$$

Factor of proportionality K [6]:

$$K = \frac{\dot{\gamma}}{\tau^n} \qquad (5.31)$$

The die constant G for a square is calculated from [6]

$$G_{\text{square}} = 0.5 \cdot \left(\frac{0.42}{3}\right)^{\frac{1}{n}} \cdot \frac{d_0^{\frac{3}{n}+1}}{2\,L} \qquad (5.32)$$

The length L for a screen is taken as $L = 2\,d$ with d as the wire diameter(Fig. 5.19). Finally, the pressure drop Δp in the screen results from [6]

$$\Delta p = \frac{q^{\frac{1}{n}}}{K^{\frac{1}{n}} \cdot G_{square}} \tag{5.33}$$

For a screen with openings having a form other than a square the formula developed by Schenkel [8] can be applied.

Practical Examples with Symbols and Units

Example 1

For the following input data it is required to calculate the pressure drop Δp in the screen:

Extruder throughput $\dot{M}$ = 454 kg/h

Melt temperature T = 232 °C

Barrel diameter of the extruder D_b = 114.3 mm

Mesh number of the screen m_n = 42

Melt density ρ_m = 0.78 g/cm^3

Solution:

From Table 5.1 the minimum sieve opening for a 42 mesh square wire-gauze d_0 = 0.354 mm and the wire diameter d = 0.247 mm. The screen diameter D_s can be taken as equal to the barrel diameter D_b. Inserting the respective values with the units above, we get from Eq. (5.25) $\dot{q}$ = 0.00576 cm^3/s and from Eq. (5.27) $\dot{\gamma}_{square}$ = 928.15 s^{-1}.

The viscosity η at $\dot{\gamma}_{square}$ = 928.15 s^{-1} is found to be 353.84 Pa s with T = 449.6 °F for the given HDPE.

The power law exponent n follows from Eq. (5.28) to be n = 1.589 for the respective resin constants.

Shear stress τ = 328414 N/m^2.

Factor of proportionality K in Eq. (5.31) K = 1.586 10^{-6}.

Die constant G:

$$G_{square} = 0.5 \cdot \left(\frac{0.42}{3}\right)^{\frac{1}{1.589}} \frac{\left(\frac{0.354}{100}\right)^{\frac{3}{1.589}+1}}{2\cdot 2\cdot \frac{0.247}{1000}} = 1.591\cdot 10^{-8}$$

Finally, the pressure drop Δp in the screen from Eq. (5.33) is Δp = 18.33 bar.

Example 2

A screen pack consists of two 42 mesh screens, two 100 mesh screens, and one 200 mesh screen (Fig. 5.20). What is the pressure drop in the screen pack for the same conditions as above?

Solution:

By applying the equations above, following values can be obtained for Δp:

for the 42 mesh: 18.3 bar

for the 100 mesh: 24 bar

and

for the 200 mesh: 26.9 bar

The total pressure drop Δp in the screen pack is therefore $\Delta p = 2 \cdot 18.3 + 2 \cdot 24 + 26.9 = 111.5$ bar. ∎

Example 3

What is the pressure drop if the screen is 70% blocked for a screen of mesh size 42 for the same input as above?

Solution:

As the diameter of the screen D_s is proportional to the square root of the area, the new diameter for the blocked screen has to be multiplied by the factor $\sqrt{(1 - 0.7)}$.

This leads to a pressure drop Δp of 24.8 bar for the same conditions as in Example 1. ∎

Results of Calculations

The effect of the type of polymer on the pressure drop in a screen pack consisting of screens of varied mesh size as shown in Fig. 5.20 is presented in Fig. 5.22. As can be expected, the pressure drop in the screen pack decreases rapidly with decreasing melt viscosity.

Figure 5.23 shows that the influence of the extruder throughput on Δp becomes less pronounced with increasing throughput. The effect of mesh size on Δp is given in Fig. 5.24. The finer gauze with a mesh of 325 has about one and a half times more resistance than a mesh of 42. The exponential increase in pressure with the reduction of screen area, when the screen is blocked by undesirable material, can be seen from Fig. 5.25.

Screen packs are used in extruders for various purposes, such as melt filtration, better back mixing of the melt, and better plastication of the resin. In all these applications it is important to be able to predict the pressure drop in the screen, as it affects the extruder throughput significantly. The design methods presented here on the basis of recent advances in computational rheology accurately simulate the influence of the polymer, mesh size, resin throughput, and screen blocking.

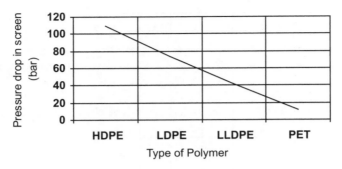

FIGURE 5.22 Effect of the polymer type on the pressure drop Δp in the screen pack

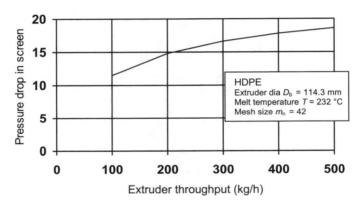

FIGURE 5.23 Pressure drop in screen vs. extruder throughput

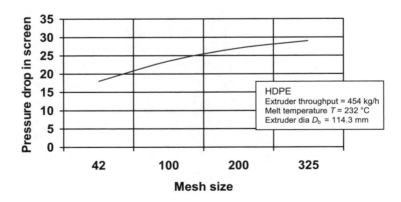

FIGURE 5.24 Effect of the mesh size on the pressure drop in the screen

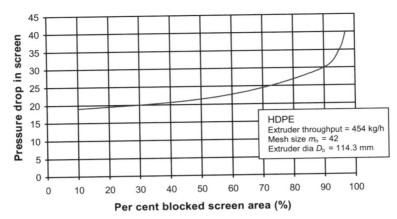

FIGURE 5.25 Effect of reduced screen area on the pressure drop in the screen

TABLE 5.2 Dimensions of Square Screens [1]

Mesh size	Sieve opening (mm)	Nominal wire diameter (mm)
42	0.354	0.247
100	0.149	0.110
200	0.074	0.053
325	0.044	0.030

■ 5.7 Parametrical Studies

Parametrical studies in polymer processing are of importance for determining the effect of significant parameters on the product quality. This section deals with those studies in pipe extrusion, blown film and thermoforming as examples, and includes relevant data on the dies and screws commonly employed.

5.7.1 Pipe Extrusion

Drawdown and Haul-Off Rates

Drawdown occurs when the velocity of the haul-off is greater than the velocity of the extrudate at the die exit (Fig. 5.27). This leads to a reduction of the extrudate cross-section. The draw ratio D_R can be expressed as (Fig. 5.28)

$$D_R = \frac{D_{die}^2 - D_{mandrel}^2}{OD_{pipe}^2 - ID_{pipe}^2} \tag{5.34}$$

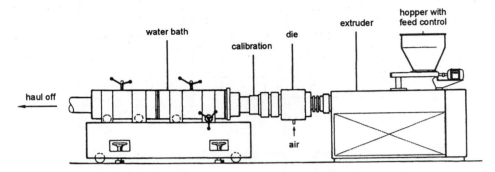

FIG. 5.26 Pipe extrusion [17]

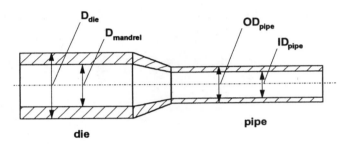

FIGURE 5.27 Draw-down in pipe extrusion

FIGURE 5.28 Relationship between draw ratio and haul-off rate [16]

The draw ratio is dependent on the resin and on the haul-off rate. In Fig. 5.28 [16] the ratio of die diameter to pipe diameter, which in practice is referred as the draw ratio, is shown as a function of the haul-off rate for a PA resin. For other resins this ratio has to be determined experimentally.

Table 5.3 shows typical data obtained on twin screw extruders for pipes. Design data for pipe dies and sizing units (Fig. 5.29) is given in Table 5.4 [21].

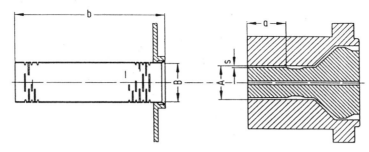

FIGURE 5.29 Design data for pipe dies and sizing units (see Table 5.3) [21]

TABLE 5.3 Typical Data for Twin-Screw Extruders for Pipes Made of Rigid PVC [21]

Screw diameter D (mm)	Screw length L/D	Screw speed (min^{-1})	Screw power (kW)	Specific energy (kWh/kg)
60/70	18/22	35/50	15/25	0.1/0.14
80/90	18/22	30/40	28/40	0.1/0.14
100/110	18/22	25/38	58/70	0.1/0.14
120/140	18/22	20/34	65/100	0.1/0.14

TABLE 5.4 Die and Calibration Unit Dimensions Based on Empirical Results [21]

Pipe raw material	Outer diameter of pipe (mm)	A (mm)	B (mm)	S in % of nominal wall thickness	a at pressure 6 to 10 (mm)	b (mm)	Haul-off speed (m/min)
PVC	20	20	20.16	4	30	100/150	20/35
	160	160	161.3	79	150	500/600	2.0/3.5
PE	20	21	21	100	20	150	25/30
	160	168	167.2	100	40/175[1]	40	1.2/2.2
PP	20	21	21	100	20	150	25/30
	160	168	167.2	100	40/175[1]	640	1.0/2.0
				S [mm]			
PA 12[1]	8	14.2	8.6	2.0	25	130	55/60
PA 12[2]	20	28.2	20.85	3.3	25/30[3]	130	12/15
PA 12[2]	22	30.0	23.0	3.3	35/50[3]	130	10/12

[1] wall thickness 1 mm
[2] wall thickness 2 mm
[3] dependent on wall thickness

5.7.2 Blown Film

The blow ratio is the ratio between the diameter of the blown film to the die lip diameter (Fig. 5.30). The effect of operating variables, such as coolant temperature in the case of water-cooled films, on some film properties, such as gloss and haze, is shown in Figs. 5.31 to 5.37 [17]. Depending on the material and the type of film, blow ratios range from 1.3 : 1 to 6 : 1 [31].

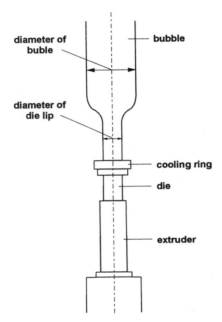

FIGURE 5.30 Blow-up in blown film

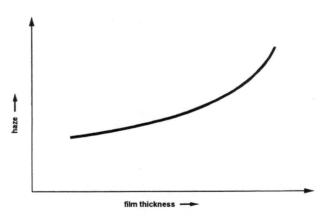

FIGURE 5.31 Effect of film thickness on haze

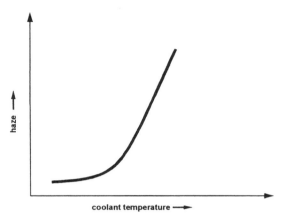

FIGURE 5.32 Effect of coolant temperature on haze

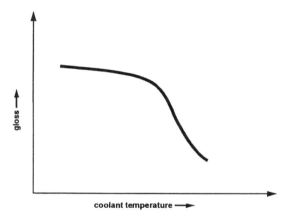

FIGURE 5.33 Effect of coolant temperature on gloss

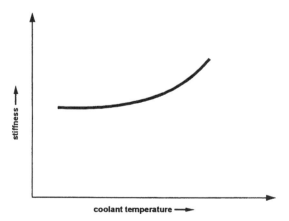

FIGURE 5.34 Effect of coolant temperature on stiffness

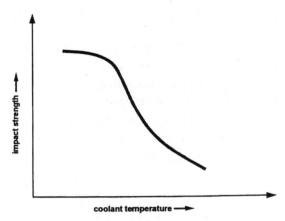

FIGURE 5.35 Effect of coolant temperature on impact strength

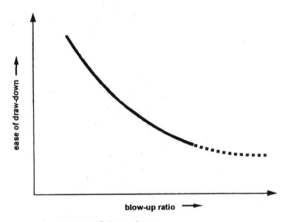

FIGURE 5.36 Effect of blow-up ratio on ease of draw-down

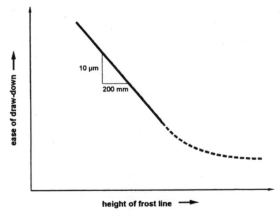

FIGURE 5.37 Effect of frost line on ease of draw-down

5.7.3 Thermoforming

In thermoforming, the material is initially in the form of a sheet or film and is shaped under vacuum or pressure after it has reached a particular temperature. At this temperature the polymer must have a sufficiently strong viscous component to allow for flow under stress and a significant elastic component to resist flow in order to enable solid shaping. Thus, the optimal conditions for thermoforming occur at a temperature corresponding to the material's transition from a solid rubbery state to a viscous liquid state [26].

The effect of different factors related to thermoforming is presented in Figs. 5.38 to 5.44 [27, 28]. Some data, including machine dependent parameters, are given in Tables 5.5 to 5.8.

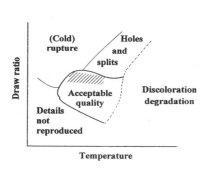

FIGURE 5.38 Parametrical relationship for thermoforming: draw ratio vs. temperature [26]

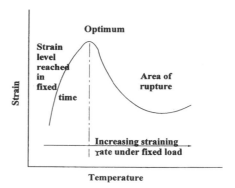

FIGURE 5.39 Parametrical relationship for thermoforming: strain vs. temperature [27]

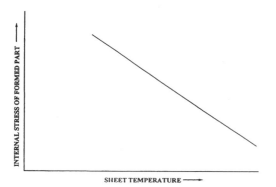

FIGURE 5.40 Parametrical relationship for thermoforming: internal stress vs. sheet temperature [27]

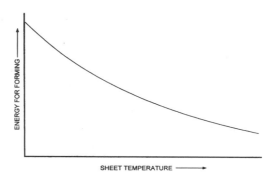

FIGURE 5.41 Parametrical relationship for thermoforming: energy vs. temperature [27]

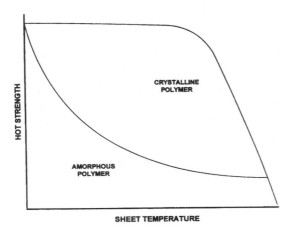

FIGURE 5.42 Parametrical relationship for thermoforming: hot strength vs. sheet temperature [28]

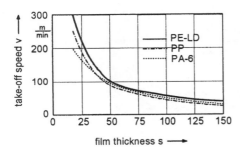

FIGURE 5.43 Production rates for flat films [23]

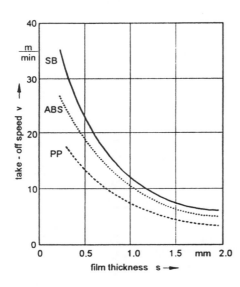

FIGURE 5.44 Production rates for thermoformable films [23]

TABLE 5.5 Guide to Thermoforming Processing Temperatures [°C] [29]

Polymer	Mold temperature [°C]	Lower processing limit [°C]	Normal forming heat [°C]	Upper limit [°C]	Set temperature [°C]
HDPE	71	127	146	166	82
ABS	82	127	163	193	93
PMMA	88	149	177	193	93
PS	85	127	146	182	93
PC	129	168	191	204	138
PVC	60	100	135	149	71
PSU	160	200	246	302	182

TABLE 5.6 Extruder Outputs for Thermoformable Film Extrusion (Film Thickness Range: 0.4 to 2.0 mm) [23]

Screw diameter D (mm)	Screw length	Output (kg/h)			
		PP	SB	ABS	PET
75	30/36 D	180/200	300/320	220/250	120/140
90	30/36 D	260/290	450/500	360/400	180/220
105	30/36 D	320/350	600/650	450/480	240/280
120	30/36 D	480/550	750/850	600/650	320/360
150	30/36 D	650/750	1100/1200	850/900	480/540

TABLE 5.7 Ranges of Melt and Roll Temperature Ranges Used in Thermoformable Film [23]

Material	Melt temperature (°C)	Chill roll temperature (°C)
PP	230/260	15/60
SB	210/230	50/90
ABS	220/240	60/100
PET	280/285	15/60

TABLE 5.8 Shrinkage Guide for Thermoformed Plastics [29]

Polymer	Shrinkage %
LDPE	1.6/3.0
HDPE	3.0/3.5
ABS	0.3/0.8
PMMA	0.2/0.8
SAN	0.5/0.6
PC	0.5/0.8
PS	0.3/0.5
PP	1.5/2.2
PVC-U	0.4/0.5
PVC-P	0.8/2.5

■ References

[1]   Procter, B.: *SPE J.*, **28** (1972) p. 34

[2]   Rao, N.: Designing Machines and Dies for Polymer Processing with Computer Programs, Hanser, Munich (1981)

[3]   Ramsteiner, F.: *Kunststoffe*, **61** (1971) 12, p. 943

[4]   Schenkel, G.: Kunststoff-Extrudiertechnik, Hanser, Munich (1963)

[5]   Squires, P. H.: *SPE J.*, **16** (1960), p. 267

[6]   Rao, N. S.: Practical Computational Rheology Primer, Proc., TAPPI PLC (2000)

[7]   Sammler, R. L.; Koopmans, R. J.; Magnus, M. A.; Bosnyak, C. P.: Proc. ANTEC 1998, p. 957 (1998)

[8]   Rosenbaum, E. E. et al.: Proc., ANTEC 1998, p. 952 (1998)

[9]   N. N.: BASF Brochure: Blow molding (1992)

[10]   Rao, N. S. and O'Brien, K.: Design Data for Plastics Engineers, Hanser, Munich (1998)

[11]   N. N.: BASF Brochure: Kunststoff Physik im Gespräch (1977)

[12]   Agassant, J. F.; Avenas, P.; Sergent, J. Ph. and Carreau, P. J.: Polymer Processing, Hanser, Munich (1991)

[13]   Rao, N. S.: Design Formulas for Plastics Engineers. Hanser, Munich (1991)

[14]   Rao, N. S.: Designing Machines and Dies for Polymer Processing with Computer Programs, Hanser, Munich (1981)

[15]   N. N.: Brochure: EMS Chemie (1992)

[16]   N. N.: Brochure: BASF AG (1992)

[17]   Perry, R. H.; Green, D.: Perry's Chemical Engineering Handbook (1984)

[18]   Carly, J. F.; Smith, W. C. : Polymer Engineering Science, p. 408 (1978)

[19]   Chung, C. I.; Lohkamp, D. T.: SPE 33, ANTEC 21 (1975), p. 363

[20]   Predöhl, W.; Reitemeyer, P.: Extrusion of pipes, profiles and cables. In: Plastics Extrusion Technology, F. Hensen (Ed.), Hanser, Munich (1988)

[21]   Winkler, G.: Extrusion of blown films. In Plastics Extrusion Technology, F. Hensen (Ed.), Hanser, Munich (1988)

[22]   Bongaerts, H.: Flat film extrusion using chill-roll casting. In: Plastics Extrusion Technology, F. Hensen (Ed.), Hanser, Munich (1988)

[23]   Michaeli, W.: Extrusion Dies, Hanser, Munich 1992

[24]   Rao, N. S.; O'Brien, K. T. and Harry, D. H.: Computer Modelling for Extrusion and Other Continuous Polymer Processes. Keith T. O'Brien (Ed.), Hanser, Munich (1992)

[25]   Eckstein, Y.; Jackson, R. L.: *Plastics Engineering*, **5** (1995)

[26]   Ogorkiewicz, R. M.: Thermoplastics – Effects of Processing. Gliffe Books Ltd, London (1969)

[27]   Titow, W. V.: PVC Technology, Elsevier Applied Science Publishers, London (1984)

[28]   Rosato, D. V.; Rosato, D. V.: Plastics Processing Data Handbook, Van Nostrand-Reinhold, New York 1990

[29]   Hermann, H.: Compound Lines in Plastics Extrusion Technology, Ed. F. Hensen, Munich 1988

[30]   Hensen, F.: Plastics Extrusion Technology, Hanser, Munich 1988

[31]   Rao, N.: *Kunststoffe*, **69** (1979), p. 126

[32]   Münstedt, H.: *Kunststoffe*, **68** (1978), p. 92

[33]   N. N.: Brochure: BASF AG (1982), p. 208

[34]   Ast,W.: *Kunststoffe*, **66** (1976), p. 186

6 Analytical Procedures for Troubleshooting Injection Molding

Injection molding has found the widest range of applications of all plastics processing technologies. Typical injection molding equipment consists of an injection unit, an injection mold, and a mold temperature control unit. The quality of the part depends on how well these components work together.

At an appropriate point in the typical molding cycle, the screw remains stationary in the forward position as indicated by point 3 in Fig. 6.1. At the end of dwell or holding pressure, the screw begins to rotate, conveying material and developing a pressure ahead of the screw. This pressure pushes the screw to move back a predetermined distance which is dependent on the desired volume of the molded part. The screw is then idle in the back position while the previously molded plastic continues cooling in the mold and while the mold opens and the part is ejected.

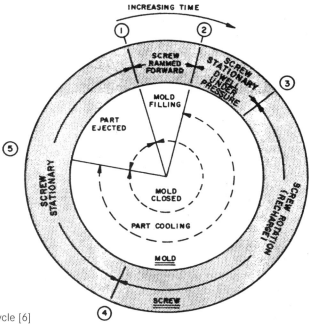

FIGURE 6.1 Molding cycle [6]

After the mold closes again, the screw is forced forward by hydraulic pressure, causing the newly recharged shot in front of the screw to flow into the empty mold. A valve, such as a check ring, prevents back flow during injection. The screw then maintains pressure on the molded plastic for a specific time called dwell or hold time, thus completing the cycle [6].

■ 6.1 Effect of Resin and Machine Parameters

6.1.1 Resin-Dependent Parameters

Injection Pressure

Injection pressure is the pressure exerted on the melt in front of the screw tip during the injection stage, with the screw acting like a plunger. It affects both the speed of the advancing screw and the process of filling the mold cavity with the polymer melt, and corresponds to the flow resistance of the melt in the nozzle, sprue, runner, and cavity. The injection pressures needed for some resins are given in Table 6.1, while the maximum injection pressure is approximately 1.2 times the pressure listed in Table 6.1 [1].

TABLE 6.1 Injection Pressure Required for Various Plastics [1]

Material	Necessary injection pressure [MPa]		
	Low viscosity heavy sections	Medium viscosity standard sections	High viscosity thin sections, small gates
ABS	80/110	100/130	130/150
POM	85/100	100/120	120/150
PE	70/100	100/120	120/150
PA	90/110	110/140	> 140
PC	100/120	120/150	< 150
PMMA	100/120	120/150	< 150
PS	80/100	100/120	120/150
PVC-U	100/120	120/150	> 150
Thermosets	100/140	140/175	175/230
Elastomers	80/100	100/120	120/150

6.1.2 Mold Shrinkage and Processing Temperature

Due to their high coefficients of thermal expansion, components molded from plastics experience significant shrinkage effects during the cooling phase. Semi-crystalline polymers tend to shrink more, owing to the greater specific volume difference between melt and solid. Shrinkage S can be defined by [2]

$$S = \frac{L^* - L}{L^*} \tag{6.1}$$

where L^* is any linear dimension of the mold and L the corresponding dimension of the part when it is at some standard temperature and pressure. As a consequence of mold shrinkage it is necessary to make the core and cavity slightly larger in dimension than the size of the finished part. Processing temperature, mold temperature, and shrinkage are given in Table 6.2 for a number of materials. The parameters influencing shrinkage behavior are shown in Fig. 6.2.

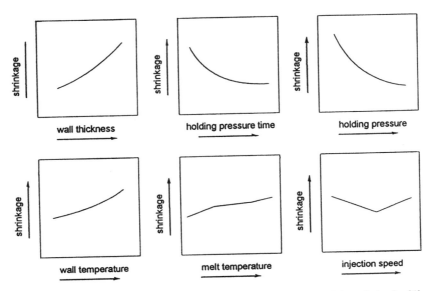

FIGURE 6.2 Qualitative relations between individual process parameters and shrinkage behavior [4]

TABLE 6.2 Processing Temperatures, Mold Temperatures, and Shrinking for the Most Common Plastics Processed by Injection Molding [1]

Material	Glass fiber content %	Processing temperature °C	Mold temperature °C	Shrinkage %
PS		180/280	10/10	0.3/0.6
HI-PS		170/260	5/75	0.5/0.6
SAN		180/270	50/80	0.5/0.7
ABS		210/275	50/90	0.4/0.7
ASA		230/260	40/90	0.4/0.6
PE-LD		160/260	50/70	1.5/5.0
PE-HD		260/300	30/70	1.5/3.0
PP		250/270	50/75	1.0/2.5
PPGR	30	260/280	50/80	0.5/1.2
IB		150/200		
PMP		280/310	70	1.5/3.0
PVC-soft		170/200	15/50	> 0.5
PVC-rigid		180/210	30/50	−0.5
PVDF		250/270	90/100	3/6
PTFE		320/360	200/230	3.5/6.0
FEP				
PMMA		210/240	50/70	0.1/0.8
POM		200/210	> 90	1.9/2.3
PPO		250/300	80/100	0.5/0.7
PPO-GR	30	280/300	80/100	< 0.7
CA		180/230	50/80	0.5
CAB		180/230	50/80	0.5
CP		180/230	50/80	0.5
PC		280/320	80/100	0.8
PC-GR	10/30	300/330	100/120	0.15/0.55
PET		260/290	140	1.2/2.0
PET-GR	20/30	260/290	140	1.2/2.0
PBT		240/260	60/80	1.5/2.5
PBT-GR	30/50	250/270	60/80	0.3/1.2
PA 6		240/260	70/120	0.5/2.2
PA 6-GR	30/50	270/290	70/120	0.3/1
PA 66		260/290	70/120	0.5/2.5
PA 66-GR	30/35	280/310	70/120	0.5/1.5
PA 11		210/250	40/80	0.5/1.5
PA 12		210/250	40/80	0.5/1.5
PSO		310/390	100/160	0.7
PPS	40	370	> 150	0.2
PUR		195/230	20/40	0.9
PF		60/80	170/190	1.2
MF		70/80	150/165	1.2/2
MPF		60/80	160/180	0.8/1.8
UP		40/60	150/170	0.5/0.8
EP	30/80	ca. 70	150/170	0.2

6.1.3 Drying Temperatures and Times

As already mentioned in Section 3.8.2, plastics absorb water at a rate depending on the relative humidity of the air in which they are stored (Table 6.3).

Moisture has an adverse effect on processing and on the part. To remove moisture, materials are dried in their solid or melt state. Table 6.4 shows the drying temperatures and times for different resins and dryers.

TABLE 6.3 Equilibrium Moisture Content of Various Plastics Stored in Air at 23 °C and Relative Humidity of 50% [5]

	CA	CAB	ABS	PA 6	PA 66	PBT	PC	PMMA	PPO	PET
Equilibrium moisture (%)	2.2	1.3	1.5	3	2.8	0.2	0.19	0.8	0.1	0.15
Water content acceptable for injection molding (%)	0.2	0.2	0.2	0.15	0.03	0.02	0.08	0.05	0.02	0.02

TABLE 6.4 Drying Temperatures and Times of Various Dryers [1]

Material	Drying temperature °C		Drying time h	
	fresh air/missed air dryers	dehumidifying dryers	fresh air/mixed air dryers	dehumidifying dryers
ABS	80	80	2/3	1/2
CA	70/80	75	1/1.5	1
CAB	70/80	75	1/1.5	1
PA 6	not recomm.	75/80	–	2
PA 66,6.10	not recomm.	75/80	–	2
PBT, PET	120	120	3/4	2/3
PC	120	120	2/4	2
PMMA	80	80	1/2	1/1.5
PPO	120	120	1/2	1/1.5
SAN	80	80	1/2	1/1.5

■ 6.2 Melting in Injection Molding Screws

In general, problems occurring during injection molding have their origin in mold design. However, with the widespread use of the reciprocating screw for conveying and plasticating, problems concerning the quality of the melt are being increasingly encountered. As in the case of plasticating extrusion, in order to achieve high part quality, the screw must be able to produce a fully plasticated homogeneous melt. Thus, the part quality is intimately connected to the plastication in the screw.

This section shows how the melting performance of the reciprocating screw can be predicted through an extension of the model of Donovan [6] to include a simplified melting model [7, 8, 9]. The resulting equations allow facile solutions to be found even with a hand-held calculator.

6.2.1 Model

During the stationary or non-rotating phase of the screw, melting of the resin takes place mainly by conduction heating from the barrel (Fig. 6.3) [3].

The parameter K characterizing the conduction melting is given by [6]

$$\frac{\left(T_m - T_b\right) \cdot \lambda_m \cdot \exp\left(-K^2 4\,\alpha_m\right)}{\sqrt{\pi \cdot \alpha_m} \cdot \operatorname{erf}\left(K / 2 \sqrt{\alpha_m}\right)} - \frac{\left(T_r - T_m\right) \cdot \lambda_s \cdot \exp\left(-K^2 / \alpha_s\right)}{\sqrt{\pi \cdot \alpha_s} \cdot \operatorname{erf} c\left(K / 2 \sqrt{\alpha_s}\right)} = \frac{-10^6 \cdot i_m \cdot \rho_m \cdot K}{2} \qquad (6.2)$$

with

$$\alpha_m = \frac{10^{-6}\,\lambda_m}{\rho_m \cdot c\,p_m} \qquad (6.3)$$

and

$$\alpha_s = \frac{10^{-6}\,\lambda_s}{\rho_s \cdot cp_s} \qquad (6.4)$$

By using the equations above, the parameter K can be obtained easily by a simple iteration. The following sample calculation for LDPE illustrates the application of Eq. (6.2).

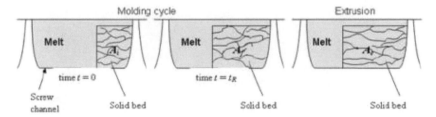

FIGURE 6.3 Change of solid bed during screw rotation in comparison to steady-state extrusion
A_i: solid bed area at the start of screw rotation
A_f: solid bed area at the end of screw rotation $t = t_R$
A_e: solid bed area for the steady-state extrusion
A_s: cross-sectional area of solid bed at any position
A_T: cross-sectional area for screw channel
A^*: solid bed ratio $A^* = A_s / A_T$

TABLE 6.5 Calculated Parameter K for Three Resins

Resin	K (m$\sqrt{s}$)
LDPE	$5.46 \cdot 10^{-4}$
PP	$1.08 \cdot 10^{-3}$
PA6	$8.19 \cdot 10^{-4}$

Sample calculation for K:

Melt thermal conductivity	λ_m	$= 0.225$ W/(m K)
Melt density	ρ_m	$= 0.78$ g/cm^3
Melt specific heat	cp_m	$= 2.4$ kJ/(kg K)
Solid thermal conductivity	λ_s	$= 0.35$ W/(m K)
Solid density	ρ_s	$= 0.92$ g/cm^3
Solid specific heat	cp_s	$= 1.9$ kJ/(kg K)
Latent heat of fusion	i_m	$= 130$ kJ/kg
Polymer melting point	T_m	$= 110$ °C
Barrel temperature	T_b	$= 200$ °C
Solid bed temperature	Tr	$= 20$ °C

In this sample calculation K was found to be $5.457 \cdot 10^{-4}$ (m/$\sqrt{s}$). For given barrel and solid bed temperatures, K is dependent only on the thermal properties of the resin. The value of K is calculated for three resins in Table 6.5.

The change of the solid bed profile is calculated from the equations derived from the model of Donovan [6] and extended to include a simplified melting model for extrusion melting. These are arranged in the following to facilitate easy iteration and calculation.

$$A_f^* - A_i^* = A_f^* \frac{\left[K \sqrt{t_T - t_R + \left(\dfrac{\delta_i}{10^4 \, K^2} \right)} - \dfrac{\delta_i}{100} \right]}{10^{-3} \, H} \tag{6.5}$$

$$A_f^* = A_e^* - \left[\frac{A_f^* - A_i^*}{1 - \exp\left(\dfrac{-\beta \cdot 2 \pi N - t_R}{60} \right)} \right] \exp\left(\frac{-\beta \cdot 2 \pi N \, t_R}{60} \right) \tag{6.6}$$

$$A_i^* = A_e^* - \left[\frac{A_f^* - A_i^*}{1 - \exp\left(\dfrac{-\beta \cdot 2 \pi N \, t_R}{60} \right)} \right] \tag{6.7}$$

The melt film thickness δ_i can be approximated by the relationship for a linear temperature profile in the melt film, and is given by [7, 8]

$$\delta_i = \delta_{av} = 0.5 \left\{ \frac{\left[2 \lambda_m \left(T_b - T_m \right) + \eta_f \, v_j^2 \cdot 10^{-4} \right] W}{10^3 \, v_{bx} \, \rho_m \left[cp_s \left(T_m - T_s \right) + i_m \right]} \right\}^{0.5}$$

(6.8)

The average temperature in the melt film is obtained from [7]

$$\overline{T_f} = 0.5 \cdot \left(T_b - T_m \right) + \frac{10^{-4} \, \eta_f \, v_j^2}{12 \, \lambda_m}$$

(6.9)

Calculation Procedure

Step 1: Calculate K using Eq. (6.2).

Step 2: Calculate δ_{av} with Eqs. (6.8) and (6.9). Substitute δ_i with δ_{av}.

Step 3: Calculate the solid bed ratio A_e^* for steady-state extrusion with the simplified model for a linear temperature profile [7].

Step 4: Find the solid bed ratio A_f^* at the end of the screw rotation using the Eqs. (6.6)

Step 5: Calculate A_i^*, the solid bed ratio at the start of screw rotation from Eq. (6.7).

The following sample calculation shows the symbols and units of the variables occurring in the equations above.

Sample Calculation

The thermal properties for LDPE and the barrel temperature are as given in the previous calculation for the parameter K. Furthermore,

Total cycle time $\qquad t_T$ = 45 s

Screw rotation time $\quad t_R$ = 22 s

Empirical parameter for all polymers:

$$\beta = 0.005$$

Screw speed $\qquad\qquad\qquad\qquad\qquad N$ = 56 rpm

Channel depth $\qquad\qquad\qquad\qquad\quad H$ = 9.8 mm

Channel width $\qquad\qquad\qquad\qquad\quad W$ = 52.61 mm

Cross-channel velocity of the melt $\quad v_{bx}$ = 5.65 cm/s

Relative velocity of the melt $\qquad\quad v_j$ = 15.37 cm/s

Solids temperature $\qquad\qquad\qquad\quad T_s$ = 20 °C

By inserting these values into Eqs. (6.5) through (6.9) and by iteration, the following target values are obtained:

Melt viscosity in the film η_f = 211 Pa s
Average temperature of the melt in the film $\overline{T}_f = 172.8\,°C$.
Average thickness of the melt film δ_{av} = $7.678 \cdot 10^{-3}$ cm.

Using $K = 5.6 \cdot 10^{-4}$, the solid bed ratios are found to be:

the solid bed ratio at the end of screw rotation, $A_f^* = 0.583$;
the solid bed ratio at the start of screw rotation, $A_i^* = 0.429$.

The solid bed ratio for steady-state extrusion, A_e^*, is calculated from the simlified melting model for extrusion [8]. Its numerical value for the conditions above is $A_e^* = 0.75$.

6.2.2 Results of Simulation

In all following figures the steady-state extrusion profile begins at the position of the stroke. The temperature of the melt refers to the temperature at the end of the screw for the case of steady-state extrusion. The solid bed ratio A^* is the ratio between the cross-sectional area of the solid bed A_s and the total cross-sectional area of the channel A_T. In Fig. 6.4 the effect of the resin type on the solid bed profiles is presented. It appears that the conductivity parameter K and the melt viscosity affect these profiles significantly, even if screw rotation and cycle times remain the same. It can be seen from Fig. 6.5 that the barrel temperature has little effect on the plastication process in the screw.

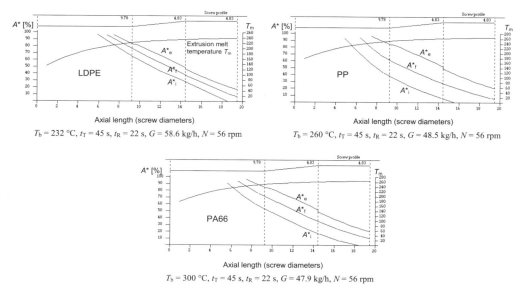

FIGURE 6.4 Effect of polymer on the melting profiles

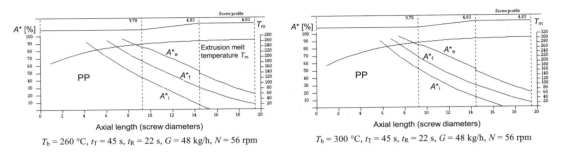

FIGURE 6.5 Effect of barrel temperature on the melting profiles for PP

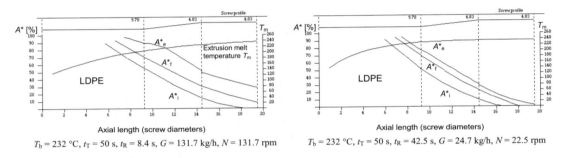

FIGURE 6.6 Effect of screw rotation time and screw speed on melting of LDPE

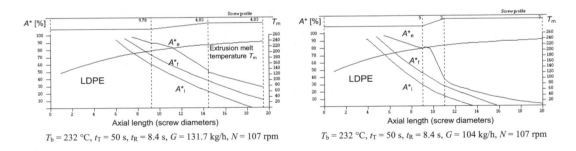

FIGURE 6.7 Effect of screw geometry on the melting for LDPE

As Fig. 6.6 depicts, slow screw speed and a high percentage of screw rotation time compared to the total cycle time strongly favor melting, which has also been found by Donovan [6]. The marked influence of screw geometry on melting becomes clear from and Fig. 6.7.

As can be expected, melting is much faster in the case of the shallower channel.

6.2.3 Screw Dimensions

Essential dimensions of injection molding screws for processing thermoplastics are given in Table 6.6 for several screw diameters [1].

TABLE 6.6 Significant Screw Dimensions for Processing Thermoplastics

Diameter (mm)	Flight depth (feed) h_F (mm)	Flight depth (metering) h_M (mm)	Flight depth ratio	Radial flight clearance (mm)
30	4.3	2.1	2 : 1	0.15
40	5.4	2.6	2.1 : 1	0.15
60	7.5	3.4	2.2 : 1	0.15
80	9.1	3.8	2.4 : 1	0.20
100	10.7	4.3	2.5 : 1	0.20
120	12	4.8	2.5 : 1	0.25
> 120	max 14	max 5.6	max 3 : 1	0.25

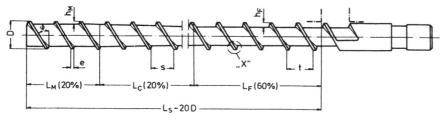

FIGURE 6.8 Dimensions of an injection molding screw [1]

6.3 Injection Mold

The problems encountered in injection molding are often related to mold design whereas in extrusion, the screw design determines the quality of the melt as discussed in Chapter 5.

The quantitative description of the important mold filling stage has been made possible by the well-known software MOLDFLOW [10]. The purpose of this section is to present practical calculation procedures that can be handled even by handheld calculators.

6.3.1 Runner Systems

The pressure drop along the gate or runner of an injection mold [15] can be calculated from the same relationships used for dimensioning extrusion dies (Chapter 5).

Calculated Example

For the following conditions, the isothermal pressure drop Δp_0 and the adiabatic pressure drop Δp are to be determined:

For polystyrene with the following viscosity constants according to [29]

A_0 = 4.4475

A_1 = -0.4983

A_2 = -0.1743

A_3 = 0.03594

A_4 = -0.002196

c_1 = 4.285

c_2 = 133.2

T_0 = 190 °C

flow rate	$\dot{m}$	= 330.4 kg/h
melt density	ρ_m	= 1.12 g/cm^3
specific heat	c_{pm}	= 1.6 kJ/(kg · K)
melt temperature	T	= 230 °C
length of the runner	L	= 101.6 mm
radius of the runner	R	= 5.08 mm

Solution:

a) Isothermal flow

$\dot{\gamma}_a$ from

$$\dot{\gamma}_a = \frac{4\,\dot{Q}}{\pi\,R^3} = \frac{4 \cdot 330.0}{3.6 \cdot \pi \cdot 1.12 \cdot 0.508^3} = 795.8\ \text{s}^{-1}$$

($\dot{Q}$ = volume flow rate cm^3/s)

a_T from

$$a_T = 10^{\frac{-c_1(T-T_0)}{c_2+(T-T_0)}} = 10^{\frac{-4.285\,(230-190)}{133.2+(230-190)}} = 10^{-0.9896} = 0.102$$

Power law exponent [29]

$n = 5.956$

η_a, viscosity [29]

$\eta_a = 132\ \text{Pa} \cdot \text{s}$

τ shear stress

$$\tau = 105013.6 \text{ Pa}$$

K:

$$K = 9.911 \cdot 10^{-28}$$

Die constant G_{circle} from Table 1.4

$$G_{\text{circle}} = \left(\frac{\pi}{4}\right)^{\frac{1}{5.956}} \cdot \frac{\left(5.08 \cdot 10^{-3}\right)^{\frac{1}{5956}+1}}{2 \cdot 0.1016} = 1.678 \cdot 10^{-3}$$

Δp_0 with $\dot{Q} = 8.194 \cdot 10^{-5} \text{ m}^3/\text{s}$ from Equation 5.1:

$$\Delta p_0 = \frac{10^{-5} \cdot \left(8.194 \cdot 10^{-5}\right)^{\frac{1}{5.956}}}{\left(9.911 \cdot 10^{-28}\right)^{\frac{1}{5.956}} \cdot 1.678 \cdot 10^{-3}} = 42 \text{ bar}$$

b) Adiabatic flow

The relationship for the ratio $\dfrac{\Delta p}{\Delta p_0}$ is [17]

$$\frac{\Delta p}{\Delta p_0} = \frac{\ln \chi_L}{\chi_L - 1}$$

where

$$\chi_L = 1 + \frac{\beta \cdot \Delta p_0}{\rho_m \cdot c_{pm}}$$

Temperature rise from Equation (5.20):

$$\Delta T = \frac{\Delta p}{10 \cdot \rho_m \cdot c_{pm}} = \frac{42}{10 \cdot 1.12 \cdot 1.6} = 2.34 \text{ K}$$

For polystyrene

$$\beta = 0.026 \text{ K}^{-1}$$
$$\chi_L = 2.34 \cdot 0.026 = 1.061$$

Finally, Δp

$$\Delta p = \Delta p_0 \frac{\ln \chi_L}{\chi_L - 1} = \frac{42 \cdot \ln 1.061}{0.061} = 40.77 \text{ bar}$$

In the adiabatic case, the pressure drop is smaller because the dissipated heat is retained in the melt.

6.3.2 Mold Filling

As already mentioned, the mold filling process is treated extensively in commercial simulation programs and by Bangert [13]. In the following sections the more transparent method of Stevenson is given with an example.

To determine the size of an injection molding machine in order to produce a given part, knowledge of the clamping force exerted by the mold is important, as this force should not exceed the clamping force of the machine.

Injection Pressure

The isothermal pressure drop for a disc-shaped cavity is given as [14]

$$\Delta p_1 = \frac{K_r}{10^5 \left(1 - n_R\right)} \left[\frac{360 \cdot \dot{Q} \cdot \left(1 + 2 \cdot n_R\right)}{N \cdot \Theta \cdot 4 \, \pi \cdot n_R \cdot r_2 \cdot b^2} \right]^{n_R} \cdot \left(\frac{r_2}{b}\right) \tag{6.10}$$

The fill time τ is defined as [14]

$$\tau = \frac{V \cdot a}{\dot{Q} \cdot b^2} \tag{6.11}$$

The Brinkman number is given by [14]

$$Br = \frac{b^2 \cdot K_r}{10^4 \cdot \lambda \cdot \left(T_M - T_W\right)} \cdot \left(\frac{\dot{Q} \cdot 360}{N \cdot \Theta \cdot 2 \, \pi \cdot b^2 \cdot r_2}\right)^{1 + n_R} \tag{6.12}$$

Calculated Example with Symbols and Units

Given data:

The part has the shape of a round disc.

The material is ABS with $n_R = 0.2565$, which is the reciprocal of the power law exponent n. The constant K_r, which corresponds to the viscosity η_p, is $K_r = 3.05 \cdot 10^4$.

Constant injection rate	$\dot{Q}$	$= 160 \text{ cm}^3/\text{s}$
Part volume	V	$= 160 \text{ cm}^3$
Half thickness of the disc	b	$= 2.1 \text{ mm}$
Radius of the disc	r_2	$= 120 \text{ mm}$
Number of gates	N	$= 1$

Inlet melt temperature T_M = 518 K

Mold temperature T_W = 323 K

Thermal conductivity of the melt λ = 0.174 W/(m·K)

Thermal diffusivity of the polymer a = 7.72 · 10^{-4} cm²/s

Melt flow angle [14] Θ = 360°

The isothermal pressure drop in the mold Δp_1 is to be determined.

Solution:

Applying Eq. (6.10) for Δp_1

$$\Delta p_1 = \frac{3.05 \cdot 10^4}{10^5 \left(1 - 0.2655\right)} \left[\frac{360 \cdot 160 \cdot \left(1 + 2 \cdot 0.2655\right)}{1 \cdot 360 \cdot 4\,\pi \cdot 12 \cdot 0.105^2} \right]^{0.2655} \cdot \left(\frac{12}{0.105}\right) = 254 \text{ bar}$$

Dimensionless fill time τ from Eq. (6.11):

$$\tau = \frac{160 \cdot 7.72 \cdot 10^{-4}}{160 \cdot 0.105^2} = 0.07$$

Brinkman number from Eq. (6.12):

$$Br = \frac{0.105^2 \cdot 3.05 \cdot 10^4}{10^4 \cdot 0.174 \cdot 195} \cdot \left(\frac{160 \cdot 360}{1 \cdot 360 \cdot 2\,\pi \cdot 0.105^2 \cdot 12}\right)^{1.2655} = 0.771$$

From the experimental results of Stevenson [14] the following empirical relation was developed to calculate the actual pressure drop in the mold

$$\ln\left(\frac{\Delta p}{\Delta p_1}\right) = 0.337 + 4.7\,\tau - 0.093\,Br - 2.6\,\tau \cdot Br \qquad (6.13)$$

The actual pressure drop Δp is therefore from Eq. (6.13)

$$\Delta p = 1.574 \cdot \Delta p_1 = 1.574 \cdot 254 = 400 \text{ bar}$$

Clamping Force

The calculation of clamping force is similar to that of the injection pressure. First, the isothermal clamping force is determined from [14]

$$F_1\left(r_2\right) = 10 \cdot \pi \cdot r_2^2 \left(\frac{1 - n_R}{3 - n_R}\right) \cdot \Delta p_1 \qquad (6.14)$$

where $F_1(r_2)$ = isothermal clamping force (N).

$F_1(r_2)$ for the example above is with Eq. (6.14)

$$F_1\left(r_2\right) = 10 \cdot \pi \cdot 12^2 \left(\frac{1-0.2655}{3-0.2655}\right) \cdot 254 = 308.64 \text{ kN}$$

The actual clamping force can be obtained from the following empirical relationship, which was developed from the results published in [14].

$$\ln\left(\frac{F}{F_1}\right) = 0.372 + 7.6\,\tau - 0.084\,Br - 3.538\,\tau\,Br \tag{6.15}$$

Hence the actual clamping force F from Eq. (6.15)

$$F = 1.91 \cdot 308.64 = 589.5 \text{ kN}$$

The above relationships are valid for a disc-shaped cavity. Other geometries of the mold cavity can be taken into account on this basis in the manner described by Stevenson [14].

■ 6.4 Flow Characteristics of Injection Molding Resins

One of the criteria for resin selection to make a given part is whether the melt is an easy flowing type or whether it exhibits significantly viscous behavior. To determine the flowability of the polymer melt, the spiral test, which consists of injecting the melt into a spiral shaped mold shown in Fig. 6.9, is used. The length of the spiral serves as a measure of the ease of flow of the melt in the mold, and enables mold and part design suited to material flow.

The parameters involved in the flow process are resin viscosity, melt temperature, mold wall temperature, axial screw speed, injection pressure, and geometry of the mold. To minimize the number of experiments required to determine the flow length, a semi-empirical model based on dimensional analysis is given in this section. The modified dimensionless numbers used in this model taking non-Newtonian melt flow into account are the Graetz number, Reynolds number, Prandtl number, Brinkman number, and Euler number. Comparison between experimental data obtained with different thermoplastic resins and the model predictions showed good agreement, confirming the applicability of the approach to any injection molding resin [28].

The experimental flow curves obtained at constant injection pressure under given melt temperature, mold temperature, and axial screw speed are given schematically in Fig. 6.10 for a resin type at various spiral depths with melt flow rate of the polymer brand as a parameter. By comparing the flow lengths with one another at any spiral depth also called wall thickness, the flowability of the resin brand in question with reference to another brand can be inferred [8, 18].

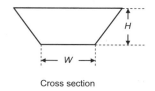

Cross section

FIGURE 6.9 Schematic representation of spiral form

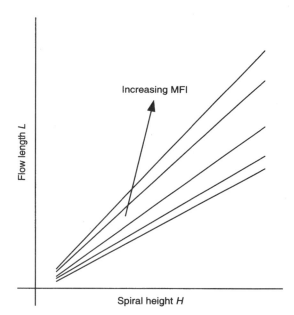

FIGURE 6.10 Schematic flow curves

6.4.1 Model

The transient heat transfer and flow processes accompanying melt flow in an injection mold can be analyzed by state-of-the-art commercial software packages. However, for simple mold geometries, such as the one used in the spiral test, it is possible to predict the melt flow behavior on the basis of dimensionless numbers and obtain formulae useful in practice. These relationships can be easily calculated with a handheld calculator offering quick estimates of the target values. Due to the nature of non-Newtonian flow, the dimensionless numbers used to describe flow and heat transfer processes of Newtonian fluids have to be modified for polymer melts.

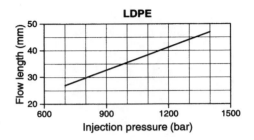

FIGURE 6.11 Effect of injection pressure on flow length

According to the authors [19, 20, 21] the movement of a melt front in a rectangular cavity can be correlated by the Graetz number, Reynolds number, Prandtl number, and Brinkman number. As the flow length in a spiral test depends significantly on the injection pressure (Fig. 6.11) the Euler number [21] is included in the present work in order to take the effect of injection pressure on the flow length.

The Graetz number Gz based on the flow length is given by

$$Gz = \frac{G \cdot c_p}{\lambda \cdot L} \tag{6.16}$$

with G melt throughput, c_p specific heat, λ thermal conductivity of the melt, and L spiral length in the spiral test.

The mean velocity of the melt front V_e in the cavity is defined by

$$V_e = \frac{Q}{A} \quad \text{with} \quad Q = \frac{G}{\rho} \tag{6.17}$$

where Q is the volume throughput, A the area of the spiral cross section Fig. 6.9 and ρ the density of the melt. The area A follows from

$$A = W \cdot H + H^2 \cdot \tan \alpha \tag{6.18}$$

with W the base width of the spiral and H spiral height or wall thickness as shown in Fig. 6.9.

6.4.2 Melt Viscosity and Power Law Exponent

Melt viscosity and power law exponent can be calculated by means of the rheological models according to Carreau, Münstedt, Klein and Ostwald and de Waele as shown by Rao [8].

The shear rate $\dot{\gamma}$ is obtained from

$$\dot{\gamma} = \frac{6 \cdot Q}{W_{\text{mean}} \cdot H^2} \quad \text{where} \quad W_{\text{mean}} = W + H \cdot \tan \alpha \tag{6.19}$$

The following dimensionless numbers are obtained from [19, 20, 21]:

Reynolds number $\qquad Re = \dfrac{V_e^{(2-n_{\text{r}})} \cdot \rho \cdot H^{\star n_{\text{R}}}}{k^\star}$ $\qquad\qquad$ (6.20)

with $H^\star = 0.5 \cdot H$ and $k^\star = \eta_{\text{a}} \cdot \dot{\gamma}^{(1-n_{\text{r}})}$

Prandtl number $\qquad Pr = \dfrac{k^\star \cdot c_{\text{p}} \cdot H^{\star(1-n_{\text{r}})}}{\lambda \cdot V_e^{(1-n_{\text{r}})}}$ $\qquad\qquad$ (6.21)

Brinkman number $\qquad Br = \dfrac{k^\star \cdot V_e^{(1+n_{\text{r}})} \cdot H^{\star(1-n_{\text{r}})}}{\lambda \cdot (T_{\text{M}} - T_{\text{W}})}$ $\qquad$ (6.22)

Euler Number $\qquad Eu = \dfrac{100 \cdot p_1}{\rho \cdot V_e^2}$ $\qquad\qquad$ (6.23)

where p_1 is the injection pressure.

6.4.3 Experimental Results and Discussion

The experimental flow curves for four different resins measured at constant injection pressure under different processing conditions and spiral wall thicknesses are given in Fig. 6.12. The flow length as a function of injection pressure is shown in Fig. 6.11 for LDPE as an example. The Graetz numbers calculated from the experimentally determined spiral lengths at different operating conditions and resins are plotted as function of the product $Re \cdot Pr \cdot Br \cdot Eu$ as shown in Fig. 6.13. As can be seen here, the correlation of the Graetz number with this product is good and thus for any particular material the spiral length can be predicted from the relationship

$$Gz = f\left(Re \cdot Pr \cdot Br \cdot Eu\right) \tag{6.24}$$

Figure 6.14 shows the good agreement between measured and calculated spiral lengths for the experimentally investigated resins.

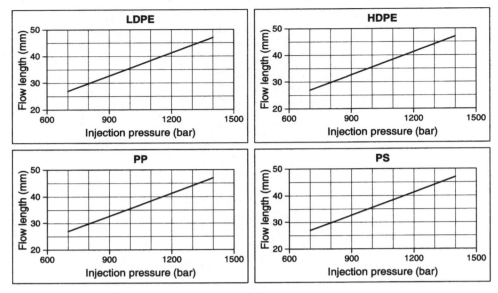

FIGURE 6.12 Experimental flow curves for LDPE, HDPE, PP, and PS

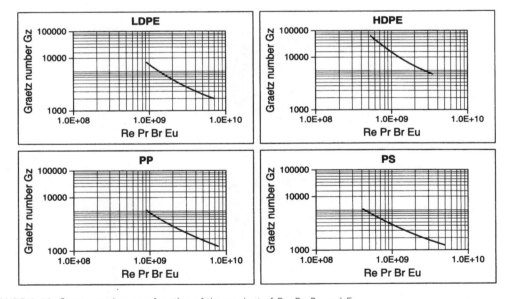

FIGURE 6.13 Graetz number as a function of the product of *Re*, *Pr*, *Br*, and *Eu*.

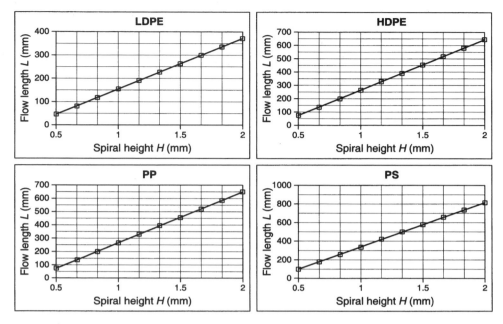

FIGURE 6.14 Comparison between measured and calculated flow length

Sample Calculation

The example given here shows how the flow length of a given resin can be calculated from Eq. (6.24). For the input values

$W = 10$ mm, $H = 2$ mm, $\rho = 1.06$ g/cm^3, $c_p = 2$ KJ/kg·K, $\lambda = 1.5$ W/K·m, $T_M = 270$ °C, $T_W = 70$ °C and $G = 211.5$ kg/h, as well as $A_0 = 4.7649$, $A_1 = -0.4743$, $A_2 = -0.2338$, $A_3 = 0.081$, $A_4 = -0.01063$, $c_1 = 4.45$, $c_2 = 146.3$ and $T_0 = 190$ °C,

following output is obtained:

$Re = 0.05964$,

$Pr = 76625.34$,

$Br = 1.7419$ and

$Eu = 10825.84$.

The Graetz number Gz for the product $Re \cdot Pr \cdot Br \cdot Eu$ follows from Fig. 6.13. $Gz = 217.63$. Hence $L = 420$ mm.

■ 6.5 Cooling of Melt in the Mold

In order to manufacture articles of high quality, mold design must take thermal, mechanical, and rheological design criteria into account. As mentioned earlier, commercial software systems [10–12] available today are of great value in accomplishing optimal mold design. But in daily practice, where quick estimates of the parameters involved are needed, the application of sophisticated software can be time consuming and costly.

This section deals with an easily applicable method of optimizing the mold design from thermal, mechanical, and rheological considerations [18].The equations involved can be solved even by hand-held calculators.

6.5.1 Thermal Design of the Mold

The cooling of the part in the mold makes up the bulk of the overall cycle time [4] and hence limits the production rate of the articles. Tempering systems for cooling the mold by cooling agents such as water are employed to achieve faster part cooling. However, these units are firstly costly and secondly, what is important, is that equally shorter cooling times can be attained, if the heat transfer between the cooling medium and the melt is optimized by a better mold design.

As mentioned in Section 3.2.1, the numerical solution of the Fourier equation, Eq. (3.31), is first presented here for crystalline and amorphous polymers [8].

Crystalline Polymers

The enthalpy temperature diagram of a crystalline polymer shows that there is a sharp enthalpy rise in the temperature region where the polymer begins to melt. This is caused by the latent heat of fusion absorbed by the polymer when it is heated and has to be taken into account when calculating cooling curves of crystalline polymers.

By defining an equivalent temperature for the latent heat (Fig. 6.15), Gloor [22] calculated the temperature of a slab using the Fourier equation for the non-steady-state heat conduction. The numerical solution of Eq. (3.34) using the correction introduced by Gloor [22] was given in [8] on the basis of the method of differences after Schmidt [27].The time interval used in this method is

$$\Delta t = \frac{c_p \cdot \rho}{\lambda} \cdot \frac{\Delta x^2}{M} \tag{6.25}$$

where

Δt =   time interval

Δx =   thickness of a layer

M =   number of layers, into which the slab is divided, beginning from the mid plane of the slab (Fig. 6.16)

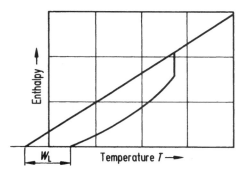

FIGURE 6.15 Representation of temperature correction for latent heat [22]

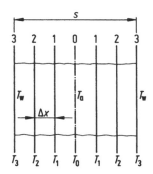

FIGURE 6.16 Nomenclature for numerical solution of non-steady state conduction in a slab [27]

The mold temperature and the thermodynamic properties of the polymer are assumed to be constant during the cooling process. The temperature at which the latent heat is evolved and the temperature correction W_L (Fig. 6.15) are obtained from the enthalpy diagram as suggested by Gloor [22]. An arbitrary difference of about 6 °C is assigned between the temperature of latent heat release at the mid plane and the temperature at the outer surface of the slab.

Figure 6.17 shows a sample plot of temperature as a function of time for a crystalline polymer.

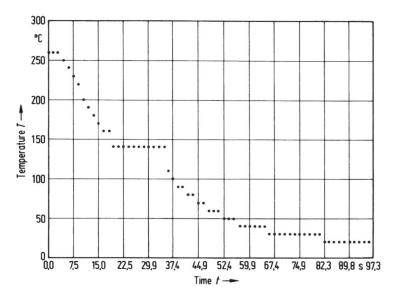

FIGURE 6.17 Plot of mid plane temperature vs. time for a crystalline polymer [8]

Amorphous Polymers

Amorphous polymers do not exhibit the sharp enthalpy change as do crystalline plastics when passing from liquid to solid. Consequently, when applying the numerical method of Schmidt [27], the correction for the latent heat can be left out in the calculation. A sample plot calculated with the computer program [18] is shown in Fig. 6.18 for amorphous polymers. It should be mentioned here that the analytical solutions for non-steady heat conduction given in Section 3.2.1 serve as good approximations for crystalline as well as for amorphous polymers.

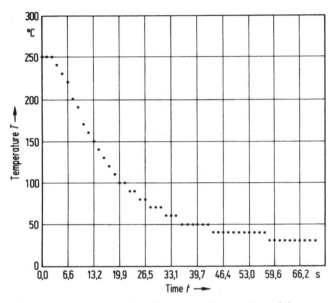

FIGURE 6.18 Plot of mid plane temperature vs. time for an amorphous polymer [4]

Calculations with Varying Mold Wall Temperature

In practice, the mold wall temperature T_w (Fig. 6.16) during part cooling is not constant because it is influenced by the heat transfer between the melt and the cooling water. Therefore, the effect of the geometry of the cooling channel layout, the thermal conductivity of the mold material, and the velocity of the cooling water on the cooling time have to be taken into account.

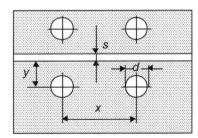

FIGURE 6.19 Schematic geometry of cooling channels

The heat transferred from the melt to the cooling media can be expressed as [23]

$$\dot{Q}_{AB} = 10^{-3} \left[\left(T_M - T_E \right) c_{p_m} + i_m \right] \rho_m \frac{s}{2} x \quad [kJ/m] \tag{6.26}$$

The heat received by the cooling agent (water) in the time t_k amounts to

$$\dot{Q}_W = 10^{-3} t_k \left(\frac{1}{\lambda_{ST} S_e} + \frac{1}{\alpha \cdot 10^{-3} 2 \pi R} \right)^{-1} \cdot \left(T_W - T_{water} \right) \quad [kJ/m] \tag{6.27}$$

The cooling time t_k can be determined from Eq. (3.38)

$$t_k = \frac{s^2}{\pi^2 a} \cdot \ln \left[\frac{4}{\pi} \left(\frac{T_M - T_W}{T_E - T_W} \right) \right] \tag{6.28}$$

with the thermal diffusivity

$$a = \frac{\lambda_m}{\rho_m c_{p_m}} \tag{6.29}$$

The influence of the cooling channel layout on heat conduction can be taken into account by the shape factor S_e according to [19, 23]

$$S_e = \frac{2 \pi}{\ln \left[\dfrac{2 x \sinh \left(\dfrac{2 \pi y}{x} \right)}{\pi d} \right]} \tag{6.30}$$

Using the values for the properties of water the heat transfer coefficient α follows from [8]

$$\alpha = \frac{0.031395}{d} \cdot Re^{0.8} \tag{6.31}$$

Re = Reynolds number.

The mold temperature T_W can now be calculated iteratively from the heat balance $\dot{Q}_{AB} = \dot{Q}_W$.

Sample calculation with Symbols and Units:

Part thickness	s	= 2 mm
Distance x	x	= 30 mm
Distance y	y	= 10 mm
Diameter of cooling channel	d	= 10 mm
Mold temperature	T_M	= 250 °C
Demolding temperature	T_E	= 90 °C
Latent heat of fusion of polymer	i_m	= 130 kJ/kg
Specific heat of the melt	c_{p_m}	= 2.5 kJ/(kg K)
Melt density	ρ_m	= 0.79 g/cm^3
Thermal conductivity of the melt	λ_m	= 0.16 W/(m K)
Kinematic viscosity of cooling water	ν	= 1.2 10^{-6} m^2/s
Velocity of cooling water	u	= 1 m/s
Temperature of cooling water	T_{water}	= 15 °C
Thermal conductivity of mold steel	λ_{ST}	= 45 W/(m K)

With the data above, the heat received from the melt is $\dot{Q}_{AB} = 12.56$ kJ/m , S_e = 3.091, Re = 8333 and α = 4300 W/(m^2 K). From the heat balance $\dot{Q}_{AB} = \dot{Q}_W = 12.56$ we obtain by iteration T_W = 37.83 °C and, finally the cooling time t_k from Eq. (6.28) to be t_k = 8.03 s.

Iteration Procedure

The iterative calculation used to obtain the mold wall temperature T_W is performed according to the termination criteria

$$\frac{\dot{Q}_{AB} - \dot{Q}_W}{\dot{Q}_{AB}} < \varepsilon ,$$

where ε is a given error bound as a tolerance. The iteration may be performed by adding a certain minimal value to the assumed mold wall temperature T_W until the termination condition is satisfied.

Results of Calculations with the Model

Figure 6.20 shows the effect of the distance y on the cooling time at two different water temperatures. It can be seen that the same cooling time can be obtained as with lower water temperature when the cooling channels are placed nearer to the mold wall. This effect can also be attained by placing the channels closer to each other, as shown in Fig. 6.21. As Fig. 6.22 depicts, the heat transfer from the melt to the coolant increases insignificantly beyond a certain coolant velocity.

Using the model and the polymers PS, LDPE, and PA as examples, cooling times were calculated and presented in Fig. 6.23 to Fig. 6.25 as functions of demolding temperatures. In these calculations, the distances x and y were kept constant at 30 mm and 12 mm, respectively. Other quantities used were u = 1 m/s, T_{water} = 20 °C and λ_{ST} = 45 W/(m·K). The resin-dependent properties were used according to the resin in question. The model also allows the simulation of the effect of thermal conductivity of the mold material as given in Eq. (6.30) (see Fig. 6.26). The effect of the temperature of cooling water on cooling time is shown in Fig. 6.27.

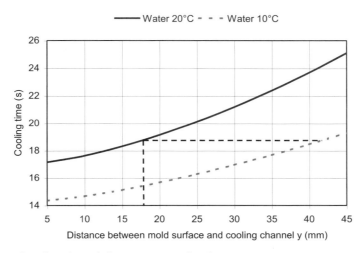

FIGURE 6.20 Effect of cooling channel distance y on cooling time

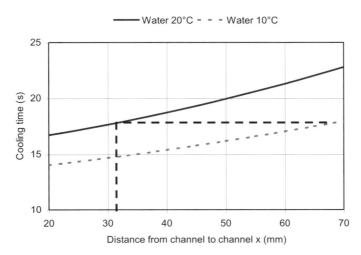

FIGURE 6.21 Effect of cooling channel distance x on cooling time

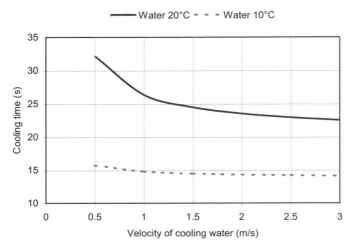

FIGURE 6.22 Influence of the velocity of cooling water on cooling time

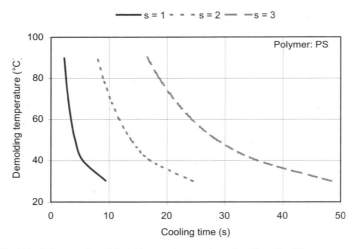

FIGURE 6.23 Relationship between demolding temperature and cooling time for PS

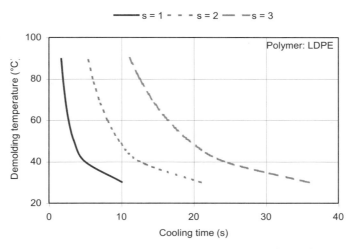

FIGURE 6.24 Relationship between demolding temperature and cooling time for LDPE

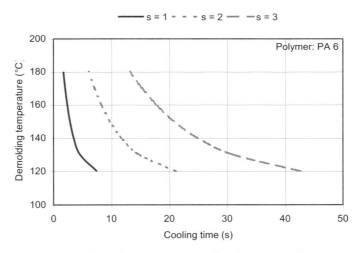

FIGURE 6.25 Relationship between demolding temperature and cooling time for PA 6

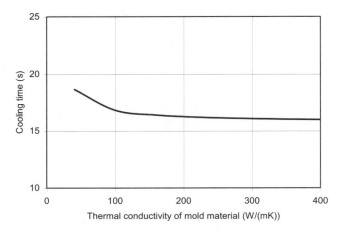

FIGURE 6.26 Effect of thermal conductivity of mold material on cooling time

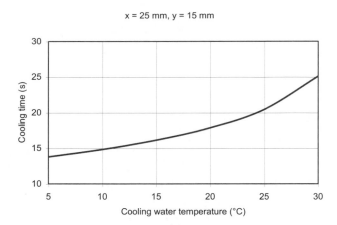

FIGURE 6.27 Influence of the temperature of cooling water on cooling time

■ 6.6 Mechanical Design of the Mold

The cooling channels should be as close to the surface of the mold as possible so that heat can flow out of the melt in the shortest time possible. However, the strength of the mold material sets a limit to the distance between the cooling channel and the mold surface. The allowable *d* in Fig. 6.28, taking the strength of the mold material into account, was calculated by Lindner [24] according to the following equations:

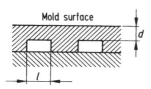

FIGURE 6.28 Geometry for the mechanical design of cooling channels

$$\sigma_{\mathrm{b}} = \frac{0.5 \cdot p \cdot d^2}{z^2} \tag{6.32}$$

$$\tau = \frac{0.75 \cdot p \cdot d}{z} \tag{6.33}$$

$$f = \frac{1000 \cdot p \cdot d^2}{z}\left(\frac{d^2}{32 \cdot E \cdot z^2} + \frac{0.15}{G}\right) \tag{6.34}$$

where

p = melt pressure (N/mm^2)

d, z = distances (mm)

E = tensile modulus (N/mm^2)

G = shear modulus (N/mm^2)

σ_{b} = tensile stress (N/mm^2)

t = shear stress (N/mm^2)

f = deflection of the mold material above the cooling channel (mm)

The minimization of the distance d can be accomplished even by a hand-held calculator [17]. The results of a sample calculation are shown in Table 6.8.

TABLE 6.8 Results of Optimization of Cooling Channel Distance

Input	Output
Melt pressure p = 4.9 N/mm^2	Channel distance z = 2.492 mm
Maximum deflection f_{max} = 2.5 µm	Deflection f = 2.487 µm
Modulus of elasticity E = 70588 N/mm^2	Tensile stress σ = 39.44 N/mm^2
Modulus of shear G = 27147 N/mm^2	Shear stress τ = 14.75 N/mm^2
Allowable tensile stress σ_{bmax} = 421.56 N/mm^2	
Allowable shear stress τ_{max} = 294.1 N/mm^2	
Channel dimension d = 10 mm	

■ 6.7 Rheological Design of the Mold

The distribution of the melt in the runner systems of multi-cavity molds [15] plays a key role in determining the quality of the part. In order to achieve this aim at low pressures, it is important to be able to predict the pressure drop of the melt flow in the relevant channels accurately. Starting with the resin rheology, the design procedure given below deals with easily applicable practical relationships [8] for calculating the pressure drop in flow channels of different geometries, commonly encountered in injection molds.

The relationship between the volumetric flow rate $\dot{Q}$ and the pressure drop Δp can be obtained by Eq. (1.10) given in Section 1.3.

The power law exponent n can be calculated from Klein's equation, Eq. (1.9), where a_1, a_{11} and a_{12} are the viscosity coefficients in the Klein viscosity model and T is the melt temperature in °F [25].

The shear rates and geometry constants for the channel shapes commonly encountered in injection molds are shown in Table 1.4.

Sample Calculation

Calculate the pressure drop of a LDPE melt flowing at 200 °C through a round channel with a length of 100 mm and a diameter of 25 mm at a mass flow rate $\dot{m} = 36$ kg/h with a melt density $\rho_m = 0.7$ g/cm^3.

Solution:

Volume flow rate:

$$\dot{Q} = \frac{\dot{m}}{\rho_m} = \frac{36}{3.6 \cdot 0.7} = 1.429 \cdot 10^{-5} \text{ m}^3/\text{s}$$

Shear rate:

$$\dot{\gamma} = \frac{4\,\dot{Q}}{\pi\,R^3} = \frac{4 \cdot 1.429 \cdot 10^{-5}}{\pi \cdot 0.0125^3} = 9.316 \text{ s}^{-1}$$

With the viscosity coefficients in the Klein polynomial [25]

$$\ln\eta = a_0 + a_1\,\ln\dot{\gamma} + a_{11}\,\ln\dot{\gamma}^2 + a_2\,T + a_{22}\,T^2 + a_{12}\,T\,\ln\dot{\gamma} \tag{6.35}$$

$a_0 = 3.3388$, $a_1 = -0.635$, $a_{11} = -0.01815$, $a_2 = -0.005975$, $a_{22} = -0.0000025$, $a_{12} = 0.0005187$, $T = 1.8 \cdot 200 + 32 = 392$ °F. η at $\dot{\gamma} = 9.316$ s^{-1} gives $\eta = 4624.5$ Pa·s

The power law exponent n follows from Eq. (1.9) to be $n = 2.052$

Wall shear stress $\tau = \eta \cdot \dot\gamma = 43077\ \mathrm{N/m^2}$

Factor of proportionality K: $K = 2.8665 \cdot 10^{-9}$

Geometry constant G_{circle} from Table 1.4:

$$G_{circle} = \left(\frac{\pi}{4}\right)^{\frac{1}{2.052}} \cdot \frac{0.0125^{\frac{3}{2.052+1}}}{2 \cdot 0.1} = 9.18 \cdot 10^{-5}$$

Finally Δp from Eq. (1.42)

$$\Delta p = \frac{\left(1.429 \cdot 10^{-5}\right)^{\frac{1}{2.052}}}{\left(2.8665 \cdot 10^{-9}\right)^{\frac{1}{2.052}}} \cdot 9.18 \cdot 10^{-5} = 689371\ \mathrm{N/m^2}$$

or $\Delta p = 6.89371$ bar

Figure 6.29 shows the calculated pressure drop Δp as function of the mass flow rate $\dot m$ of the melt at 200 °C for a viscous LDPE flowing through a round channel with the tube radius as a parameter. The appropriate runner diameter can be easily found from similar diagrams for any resin by applying the equations above.

The effect of melt viscosity on the pressure drop is given in Fig. 6.30. As can be expected, the pressure drop for the easy flowing PET is much lower than that of LDPE.

The calculation procedure allows a quick and easy prediction of the influence of the melt temperature, as depicted in Fig. 6.31.

In addition, the effect of the geometry of the runner cross section can also be easily simulated by means of the relationships given in Table 1.4, as shown in Fig. 6.32 for a square cross-sectional area compared to a circle for the same cross-sectional area under otherwise equal conditions.

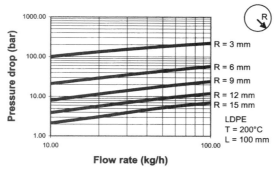

FIGURE 6.29 Pressure drop vs. flow rate for a circular cross-section for LDPE

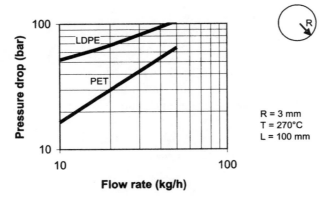

FIGURE 6.30 Effect of melt viscosity on pressure drop

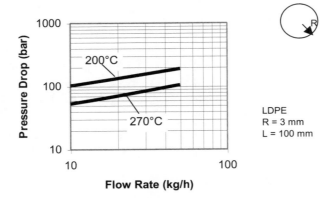

FIGURE 6.31 Effect of melt temperature on pressure drop

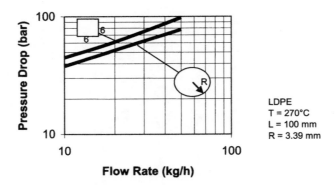

FIGURE 6.32 Effect of channel shape on pressure drop

General Channel Shape

For channel shapes other than those given in Table 1.4, the equation of Schenkel [26] can be used with good accuracy.

For a circle with $n = 1$, Eq. (5.12) leads correctly to R_{rh} = radius of the circle. Using this equation, the pressure drop is calculated for the shape shown in Fig. 6.9 and plotted over the flow rate. In this case, first the substitute radius is obtained, and then the same calculation procedure as for the round channel applied.

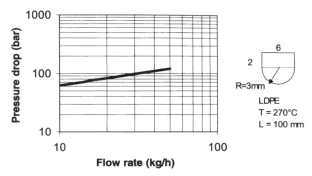

FIGURE 6.33 Pressure drop for a noncircular channel with R_{rh} = 2.77 mm and n = 2.052

■ References

[1]   Johannaber, F. : Injection Molding Machines 4th ed., Hanser, Munich (2007)
[2]   Mckelvey, J. M.: Polymer Processing, John Wiley, New York (1962)
[3]   Poetsch, G.; Michaeli, W.: Injection Molding – An Introduction, Hanser, Munich (2007)
[4]   Zoellner, O.; Sagenschneider, U.: *Kunststoffe*, **84** (1994) 8,p. 1022
[5]   Hensen, F. (Ed.): Plastics Extrusion Technology, Hanser, Munich (1988)
[6]   Donovan, R. C.: *Polymer Eng. Science*, **11** (1971), p. 353 and p. 361
[7]   Tadmor, Z. and Klein, I.: Engineering Principles of Plasticating Extrusion, Van Nostrand Reinhold, New York (1970)
[8]   Rao, N. S.: Design Formulas for Plastics Engineers, Hanser, Munich (1991)
[9]   Rao, N. S. et al.: Proceedings of SPE ANTEC 2001
[10] N. N.: C-Mold, Advanced CAE Technology Inc., U. S. A.
[11] Austin, C.: CAE for Injection Molding, Hanser, Munich (1983)
[12] N. N.: CADMOULD, IKV Aachen, Germany
[13] Bangert, H.: Doctoral Thesis RWTH Aachen (1981)
[14] Stevenson, J. F.: *Polymer Eng. Science*, **18** (1978), p. 573
[15] Paulson, D. C.: Pressure Losses in the Injection Mold, *Modern Plastics* (1967), p. 119
[16] N. N.: Brochure Mannesman-Demag 1992
[17] Rao, N. S. and O'Brien, K.: Design Data for Plastics Engineers, Hanser, Munich (1998)
[18] Rao, N. S. et al.: Proceedings of SPE ANTEC 2003

[19] Throne, J. F.: Plastics Process Engineering, Marcel Dekker Inc., New York (1979)
[20] Kamal, M. R.; Kennig, S.: *Polymer Eng. Science*, **12** (1972)
[21] Perry, R. H.; Green, D.: Perry's Chemical Engineering Handbook (1984)
[21] Gloor, W. E.: Heat Transfer Calculations,Technical Papers Volume IX-C
[23] N. N.: VDI-Wärmeatlas, VDI-Verlag, Düsseldorf (1984)
[24] Lindner, E.: Designing Injection Molds (German), VDI-Verlag, Duesseldorf (1974), p. 109
[25] Klein I.; Marschall, D. I.; Friehe, C. A.: *SPE J.*, **21** (1965), p. 1299
[26] Schenkel, G.: Private Communication
[27] Schmidt, E.: Einfuerung in die Technische Thermodynamik, Springer, Berlin (1962)
[28] Rao, N.: *Kunststoffe*, **73** (1983) p. 696
[29] Münstedt, H.: *Kunststoffe*, **68** (1978) p. 92

Summary

Polymer processing machinery can be designed using sophisticated software available today on the market. However, in daily practice, when quick estimates of the parameters involved are needed, the application of elaborate software can be time consuming and costly.

Starting from resin rheology, thermal properties of polymers in solid and melt state and heat transfer to polymers, this book presents analytical procedures for quick troubleshooting of polymer processes and machinery based on simple, proven design formulas.

The methodology for troubleshooting as applied in this book lies in examining the compatibility between machine design and the desired target process parameters. This means that if a machine element such as a screw or a mold does not lead to the target variables desired, its design has to be changed and suited to the resin and process parameters. The important design tools necessary to perform this task enable the plastics engineer to address the day-to-day problems occurring on the shop floor quickly and easily.

Detailed case studies and worked-out practical examples illustrate this approach. The areas covered include extruder screw and die design, blown film, blow molding, pipe extrusion, and thermoforming, to mention a few.

Cooling of the melt in injection molds, rheological design of runner systems, designing injection molding screws and melt flow in spiral injection molds are some topics among many dealt with in the area of injection molding.

This book is not an exhaustive work. The authors still hope that it will be a good source of help for solving day to day problems occurring on the shop floor of the plastics industry using practical analytical techniques. The Appendix contains a list of simple, but proven computer programs.

Appendix:
List of Programs
with Brief Descriptions

VISRHEO

This program calculates the viscosity coefficients occurring in four different viscosity models commonly used, namely Carreau, Münstedt, Klein, and Power law on the basis of measured flow curves and stores these coefficients automatically in a resin data bank.

TEMPMELT

This program calculates the solids melting profile of a multi-zone screw taking non-Newtonian melt flow fully into account. A resin data bank containing thermal properties and viscosity constants of different models is part of this program. Melt temperature, pressure and motor horse power are also calculated.

VISSCALE

This program is used to scale up multizone single screws. The resin data bank is included.

VIOUTPUT

This program determines the output of a single screw extruder for a given screw speed or screw speed for a given output. The resin data bank is part of the program.

VISPIDER

Program for designing spider dies including resin data bank.

VISPGAP

Program for designing spiral mandrel dies including resin data bank.

VISCOAT

Program for designing coat-hanger type dies, including resin data bank.

All programs above run on a PC in a user friendly dialogue mode under Windows XP or Vista. These programs can be obtained from Natti S. Rao by contacting his e-mail address raonatti@t-online.de

Index

A

absorption   60
absorptivity   54
apparent shear rate   2
apparent viscosity   3

B

barrel temperature   162
Biot number   48, 49
blow molding dies   132
blown film   146
Brinkman number   48, 49
buckling   108

C

capacitance   57
clamping force   167
coating   61
conduction, hollow sphere   31
convection   51
cooling   174
cooling channel   182

D

Deborah number   48, 49
deflection   107
desorption   60
die constant   113
dielectric heat loss   57
dielectric loss factor   57
dissipation   47
drawdown   143
draw ratio   143
drying   157

E

enthalpy   20
equivalent radius   114
extrusion   67
extrusion dies   111

F

feed zone   68, 69
Fick's law   59
flat dies   135
flow rate   8
Fourier number   38, 48, 49

G

gate   163
geometry constant   10
Graetz number   48, 49
Grashof number   48, 50
Griffith number   50

H

haul-off   143
heat deflection   25
heat penetration   24
heat transfer   29

I

injection mold   163
injection molding   153
injection molding screws   157
injection pressure   154, 166

K

Klein model   4

L

laminating   61
Lewis number   48, 50

M

melt film   83
Melt Flow Index   7
melt fracture   130
melting parameter   87
melting profile   87
melting rate   86
melt viscosity   170, 186
Melt Volume Index   7
mesh size   137
metering zone   68, 71
moisture   157
mold filling   166

N

Nahme number   48, 50
Nusselt number   48, 50

P

Peclet number   48, 50
pelletizer dies   132
permeability   59, 62
power law   3
Prandtl number   48, 50
pressure drop   112, 142, 184

R

radiation   53
reflectivity   54
residence time   123, 126
Reynolds number   48, 50
runner   163

S

Schmidt number   48, 50
screen   137
screw power   95
screw rotation   162
shear rate   10, 114, 126
Sherwood number   48, 50
shrinkage   15, 155
specific heat   18
specific volume   15
spider dies   124
spiral dies   129
steady state conduction   29
Stefan-Boltzmann constant   53
Stokes number   48, 50

T

temperature distribution   38
thermal conductivity   22
thermal diffusivity   23
thermal expansion   19
thermoforming   149
three-zone screw   68
torsion   107
transition zone   68
transmissivity   54

U

unsteady state conduction   37

V

Vicat softening point   26
volumetric flow rate   184

Biography

Natti S. Rao obtained his B. Tech (Hons) in Mechanical Engineering and M. Tech in Chemical Engineering from the Indian Institute of Technology Kharagpur, India. After receiving his Ph. D. in Chemical Engineering from the University of Karlsruhe, Germany, he worked for long years as a senior research scientist with the BASF AG. Later he served as a technical advisor to the leading machine and resin manufacturers in various countries.

Dr. Rao has published over 70 papers and authored five books on designing polymer machinery with the help of computers. Prior to starting his consulting company in 1987, he worked as a visiting professor at the Indian Institute of Technology Madras. Besides consulting, he holds seminars on computer aided design of polymer machinery. Dr. Rao gave lectures on polymer engineering at the University of Texas Austin, and is presently involved in the continuing education of UMASS Lowell, USA.

Dr. Rao is a Fellow of the Society of Plastics Engineers USA.

Nick R. Schott received his B. S. in Chemical Engineering in 1965 from the University of Berkeley and MS and Ph. D. from the University of Arizona in 1971. Prof. Schott taught Plastics Engineering 39 years at the UMASS Lowell, and became an Emeritus professor in 2010. He was advisor and co-advisor for over 100 MS and doctoral students. Nick presented numerous articles in the SPE ANTEC Proceedings.

Prof. Schott is a Fellow of the Society of Plastics Engineers USA.

From Structure to Flow Behavior and Back Again.

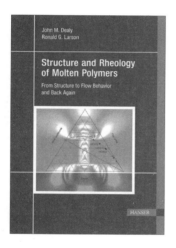

John M. Dealy/Ronald G. Larson
Structure and Rheology of Molten Polymers
From Structure to Flow Behavior and Back Again
530 pages, 130 figures, 12 tables
ISBN 978-3-446-21771-3

In recent years, several developments have made it possible to predict the detailed molecular structure of a polymer based on polymerization conditions and to use this knowledge of the structure to predict rheological properties. In addition, new techniques for using rheological data to infer molecular structure have also been developed. Soon, it will be possible to use this new knowledge to design a molecular structure having prescribed processability and end-product properties, to specify the catalyst and reaction conditions necessary to produce a polymer having this structure, and to use rheology to verify that the structure desired has, in fact, been produced.

This book provides a detailed summary of state-of-art methods for measuring rheological properties and relating them to molecular structure.

A Real Understanding of Materials, Processing and Modeling.

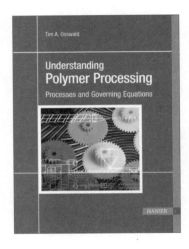

Tim A. Osswald
Understanding Polymer Processing
Processes and Governing Equations
304 pages, 266 coloured figures, 15 tables
ISBN 978-3-446-42404-3

This book provides the background needed to understand not only the wide field of polymer processing, but also the emerging technologies associated with the plastics industry in the 21st Century. The book combines practical engineering concepts with modeling of realistic polymer processes.

It is divided into three sections that provide the reader sufficient knowledge of polymer materials, polymer processing, and modeling. Understanding Polymer Processing is intended for the person who is entering the plastics manufacturing industry and as a textbook for students taking an introductory course in polymer processing. This three-part book also serves as a guide to the practicing engineer when choosing a process, determining important parameters and factors during the early stages of process design, and when optimizing such a process. Practical examples illustrating basic concepts are presented throughout the book.

More Information on Plastics Books and Magazines:
www.kunststoffe-international.com or **www.hanserpublications.com**